Slip Through a Wall

How One Can Change His Reality

By Steve Preston

1

2nd Edition

Table of Contents

Introduction

Have you ever looked at something and it made no sense? I looked at photons going through a pane of glass and it made me mad. "I can't seem to be able to walk through a wall when almost all of it is nothingness." This same light slows down in vibration and becomes radio waves and blasts it way through a wall without issue. I try to slow down and, BAM! The wall seems just as hard. Instead of slowing, it speeds to become x-rays and blasts through wall after wall. I run as fast as I can and---I stopped before I got to the wall as it looks just as solid when I'm going faster.

It doesn't make sense to me nor should it to you. How can we easily walk through water or air, but something like a wall and we seem to be in trouble?

This book investigates the causes and possible methods to eliminate the causes of this horrible limitation we seem to have. Maybe we aren't vibrating correctly.

7

With respect to this issue, current investigators have gotten wood to go through metal without the metal "seeing" it and we understand so much more about matter in general today. I think we are getting very close to the answer. One thing we know is that besides going through a window, if light goes the right frequency it goes through the wall in the form of Cosmic Rays wreaking havoc and death as it goes. The odd part is that as Aether and fermions, the building blocks of all matter, continue to vibrate faster, the characteristic changes dramatically and all of a sudden, the waves associated with the super high frequency vibration turn solid as some type of magnetic mass and is again halted by the wall. [I'm going to go over this a little later.]

Man-oh-man-oh-man! All I want to do is go through a wall. I don't need anomalies springing up all over the place.

If you read my books on "Vibrational Matter" and "Anthropic Reality" you might be one step closer to pushing yourself through a wall, but let me see if I can get you a little closer. The final steps will have to be on your own because, I'm not sure I was actually able to go through a portion of a wall and if I did it was only slightly and for an instant.

Step out of you comfort zone for a little bit and read this book. I think you will see walls a little differently afterwards. Speaking of being different, let's first change the concept of matter. Think of matter as being a vibrational field. That is exactly what researchers are finding today. I'm not talking about wimpy guys either; I'm talking about almost all the great physicists; Neils Bohr, Albert Einstein, Milo Wolff, and many others. Once we get a newer understanding of what a wall or any "matter" is, I'll need to introduce you some details of science during ancient times before being you into the present and establishing various methods that can be employed to slip through a wall. I'm not saying it's going to be easy, but there are methods. Some won't seem like they are part of our "Reality" so we will have to look at our reality, what makes up matter in our reality and something called Participative Anthropics. If you have heard about Schrodinger and his cat in a box, you have been introduced to it, but we will need to learn more to get through that wall.

Vibrations Make Matter

To start off there is a major change in the way you should view stuff, on the following page is a table that shows the actual or theoretical frequency and wavelength standards of common elements known today as categorized under what is commonly known as the Tamashii Atomic Model. The frequencies have been derived from the various groups investigating the Tamashii model but actual manufacture of some of these frequencies has been somewhat difficult. How would you like some particles vibrating at 60 exahertz? That vibration causes Gold, as you can see from the list following. Have the right frequency and make the material you want. Some believe light and matter are the same and vibrating frequencies that create gamma rays may also start to create matter. In this case it is the tiniest block of matter we call hydrogen. Most today believe there is a significant separation between matter and electromagnetic force, so vibrational levels of various elements of matter are contained in the following chart.

Brief Chart of Vibration Definitions

Name /atomic #	Maximum Wavelength [meters]	Highest Frequency [Hertz]
Aether		0
fermion	1×10^{8}	20×10^{5}
Hydrogen/1	1×10^{9}	30×10^{16}
Beryllium/9	1×10^{10}	30×10^{17}
Silicon/28	3.5×10^{11}	8.5×10^{18}
Zirconium/91	1×10^{11}	30×10^{18}
Gold/197	5×10^{12}	60×10^{18}
Meitnerium/270	3.7×10^{12}	27×10^{19}
Black Holes	'0	Very fast

Matter, according to Einstein and a raft of others, comes from something called the potential for matter or "Aether". At this state it is "essentially" nothing so one can easily walk through it and even when it becomes a mass of fermion "sub-particles" there is no resistance in moving through it. Hydrogen and many types of "gas matter' are easily pushed out of the way, but soon, we are stopped; not by the rigid construct of mass, but by the vibration level of the invisible Aether. To make this even stranger, Anthropics, Quantum Mechanics, and multi-verse timelines of reality have completely changed how we view everything. It seems reality is manipulated by what some call our souls. If you are a religious person and wondered WHY God Incarnate told his followers that *if they had enough faith they could move the positions of mountains*, the answer is simple; they could and we can today. If we

11

can change reality itself, going through a wall will be child's play.

People are really made as three entities like the Bible indicates we are made in the image of the three entity creator God, we have a "self" [the part we see and we interact with our with], the" Soul" [that little voice that gives you insight while you are sleeping and during times of stress, and finally the "spirit" [This part allows us to transfer between universes and is not really part of this discussion.] We will see that when it comes to getting your "soul" to have more influence in "reality", you must vibrate faster. I know you are wondering how to do this, but before we go on, have you heard about this Theory of Relativity that Einstein came up with? He told us that if we travel in a circle [spin] or in a straight line at near the speed of light, we will never age. Odd as that sounds, it has been shown to be correct. While your body in your reality would be getting older, in the "reality" everyone else is experiencing, doesn't even see you age at all. When you finally stop spinning, you start aging again. If you looked at a wall before you started spinning, it would disappear as you got fast enough.

Our first method to walk through a wall is to simply spin in a circle fast enough until it disappears. Make sure you don't stop spinning until you get to the other side.

How Can you Go So Fast?

If you were worried about it you don't have to run the speed of light for all this to happen as vibrating is just as effective, so how can you vibrate fast enough? This is not so easy.

These vibrations need a catalyst to change. Instead of the super high frequencies associated with dense matter, these catalysts vibrate at extremely low frequencies. We can sense how these catalyst frequencies affect us by looking at our brain. When it comes to brain functions, everyone seems to want to experiment with catalyst vibrations. The list below is a tiny segment of the discoveries made when electromagnetic or pressure waves bombard the brain at extremely low frequencies.

Hertz	Called	Condition Noted
0.2	Epsilon	Out of Body Experience
0.30	Epsilon	Depression
0.40	Epsilon	Confusion
0.50	Delta	Relaxing & pain relief
1.00	Delta	Feeling of well-being
6.40	Theta	Accelerated learning
7.83	Theta	Stress tolerance
8.6	Alpha	Induced sleep
9.5	Alpha	Positive Thinking
10	Alpha	Epileptic Seizure
12	Beta	Inc. Mental Ability
14	Beta	Injury Recovery
15	Beta	Euphoria

One potential reason for brain function to be affected by various frequencies is that the brain actually can be attuned to these various frequencies. One might wonder if the brain could be focused on a particular frequency that could change the dynamics of matter nearby. My brain is too weak for such a thing, so I cheated with something called "Infratonic Subtraction". The method takes two almost identical sounds and one ear hears one sound while the other hears the second one. Miraculously, the brain subtracts the 2 sounds [like noise cancelling headphones] to make an <u>extremely low frequency</u> you cannot hear, but these vibrations act on the brain to allow interesting, almost mystical, feelings and visions. Even with that, I had very limited results, but I will go over that experience some later. As you read, please keep it in the back of your mind the possibility of actually changing your immediate "reality", including that wall in front of you. I'll also go over some of the interesting results of the experimentations that have been conducted over the years.

During ancient times texts told about magic "vibrating" crystals. As walking through a wall may have something to do with vibrating let's look into these ancient mystics.

Magic Crystal

From the introduction, you probably are thinking I'm a nutcase, but please stay with me a while, and you will gain a broader appreciation of how our universe and how "YOU" work. First thing to be sensitized to is that everything is vibrational. Particles are simple vibrating nodes, light is simple vibrating fields, and even you are simple a vibrating life force. With all this vibrating, there should be things that can vibrate. That is where crystallized structures of quartz and other homogeneous materials come in. You have probably heard about crystals having some magical power and dismissed it as some type of belief destined to go along with astrology and extracts of poppy seeds. The Tamashii model of atomic structure and this whole concept of vibrating particles may give credence to the notion that crystals hold magic. If you start with a crystal of a homogeneous material that is locked in a covalent lattice structure, it will tend to vibrate at a very specific frequency when excited and the vibrations will continue for some time due to the resonance of the crystalline substrate. In other words, a crystal could cause a continuing vibration. A secondary vibration from a sound cue or other stimulus could very well produce "beat frequencies" or frequencies that add together to form higher frequencies, which possibly could be high enough to modify the atomic characteristics of a material in close association with the crystal. All this seems like hocus pocus, but that is exactly what causes your crystal watch to work and how a transistor amplifies a signal in your television set. The concept is only modified because the output desired is this extremely high frequency vibration pattern that affects particles. Before I go on, let me show you a couple of images from the Crystal Cave in Mexico. The little person near the center of each picture is a real person.

Crystals Make Light

If a crystalline lattice is bombarded with vibrational energy, the energy is stored in the atomic clouds and electrons spinning around are pushed away from their associated nucleus, temporarily. Soon, the nuclear forces drive the electrons back into place. This motion has a characteristic energy flow that forces the emission of a photon [the mass characterization of light] or it makes the photonic distinction turn visible as a particular vibrational mode. All

16

the newly established photons in the crystalline lattice react the same way because all the electrons were moved out similarly so a massive beam of light is generated at "generally" the same frequency. Today we call this action, **L**ight **A**mplification by the **S**timulated **E**mission of **R**adiation or **<u>LASER</u>** emission. This laser phenomenon also is experienced when other atomic elements are excited and return to their assigned spaces. All this back and forth looks like vibrations and they are. These vibrations do a whole lot more than produce light, but later we will see that making light is an important key element in walking through a wall.

Caution

Don't discount the magic crystal thing, but don't go out and get a crystal to make you feel better either. It probably will just sit there and do nothing for you. Just open your mind to possibilities that ancient humans could do marvelous things that we are only now beginning to understand. Because they were talented in some exotic ways the ancient people decided they should make electricity. Let's investigate a place where crystals could really be compressed to make electricity, magnetism and light.

Vibrational Electricity

There are many ways to put energy into an atomic cloud so you can get it to vibrate differently. One is by agitating it with a secondary electromagnetic field. The second way is by compressing the lattice structure of a crystal. When electromagnetic fields are made this way it is called piezo-electricity. The early Kodak Flash cubes made light this way as an anvil would simply smash a crystal so hard that a substantial amount of light was produced. Smashing produced light and releasing the pressure also made light and electricity. The image below shows a number of the Kodak "magic cube" flash lamps.

Unlike typical flash cubes that require electricity, these things created their own electricity when a small crystalline stone was compressed and released. This is important as we go on. I decided to see how these things worked long ago and tore it apart. There was a stone in the shape of a cube. While it did not look like anything, I took the cube and went into a closet so it would be dark and carried the cube in a vice. By turning the vice with enough pressure, I could hear a pop and a flash of light. Releasing the cube made another pop and another flash of light as electricity was "magically" generated. Guess what!!!! You could do this pressure and release over and over again and the cube never ran out of electricity!!!!!! Someone discovered this crazy property long ago. The stone was pyro-electric and produced a flame as electricity was generated from the crystalline structure being changed by the pressure of the vice. During ancient times, people needed electricity and there is ample proof that people made electricity out of this "piezo-electric effect" of crystals. To make this even odder, these ancient scientists manufactured alternating current just like we use today. It seems that one of the major developments from before the end of the Pleistocene, 10 thousand years ago was an electric generating plant. I know it will be hard to believe but stay with me for a little before rejecting this notion. Now for the nuts part, we call the electric plant "the Great Pyramid" and we can be pretty sure that it was working before the worldwide flood that marked the Pleistocene Extinction and it was still operational for thousand years after the flood. For this study, all you need to know is that there are 5 huge high crystalline, critically polished, red granite slabs that completely cover the highest chamber of the pyramid. These oscillating crystals were pressurized by the weight of the pyramid, just like the Kodak flash cubes, to make light and electric discharges. When compressed the light produced was so intense that I would ignite materials in the chamber and force the pyramid to momentarily expand and relieve pressure on the slabs. Did I tell you that releasing a pressurized crystal also made light and electricity? There was another explosion and more electricity. The crystals were then crushed again to make more electricity and another explosion, more material, another explosion and the energy from the light was collected as Electricity. The compression caused electricity going in one direction and releasing the pressure produced electricity going in the opposite direction just like Nikolas Tesla's "Alternating Current" we use today.

The diagram following shows how the pressurized crystals were placed in the explosion chamber. Notice that all were polished [even though the top 4 could not be seen at all]. This polishing was to change the resonant frequency of the giant crystalline blocks. By making them resonate at the same frequency, the flash of light was much more pronounced. These polished pyro-electric blocks would be the beginning of a massive electricity producing machine.

Once the chambers, shafts, resonating crystals and properly sized resonance cavities were in place, all the Egyptians had to do, was find a way to continuously compress the crystals enough to produce electricity. Someone thought about it for a little and might have said, "The pyramid will compress the granite by itself." Sure enough, the weight of the pyramid pushing on the granite caused electric energy to be output from the crystal. It wasn't some magic and they could not have kept the crystals from producing the initial surge of electricity, even if they wanted to. If you haven't wondered why Egyptians would have placed 5 ceiling in the room in the middle of the pyramid, let me tell you what it was not. Some officials say the 5 ceiling made the room more secure [they do not and none of the other rooms has this seemingly idiotic, multi-ceiling, construction]. When asked why all the ceilings are polished smooth even though they cannot be seen, they just shrug their shoulders. The reason people polish crystals is to set their vibrational frequency. Let's just say the pyramid made AC electricity similar to what we have today, but much higher in frequency.

Electricity Was Now Gone

From pressure on their structure, the crystal slabs, like all crystals would resonate and begin to compress and expand itself at its <u>base frequency</u>. Typically, these "oscillations" die down very quickly, but simply having this great pressure on the blocks would produce several "cycles" of electricity. Of course the electricity had nowhere to go so a very high "voltage level" would be reached and like miniature lightning bolts, electrical sparks or photon emissions would be generated. Then the energy would dissipate as heat and go away. I know this sounds like it was useless even with resonance, but the 5 polished slabs were not placed there for fun.

Resonators

This "millisecond group of electrical sparks" would have been amplified if something called "resonance" was added in the Pyramid. Resonance is a property that amplifies or sustains one frequency of electromagnetic energy while ignoring other frequencies. This property of resonance was initiated by first having the granite slabs polished to the same basic dimensions and was further extended by making the "room" directly below the granite crystals to specific dimensions. Finally, the resonance was enhanced by building sort of a tuning fork or vibration cavity at the center of the room. Some may tell you that this vibration cavity is a sarcophagus, but it doesn't make sense. There was never a lid and no evidence that a body had ever been placed in the tuning fork. By the way, the frequency that all of these elements seem to amplify is 640 Hertz.

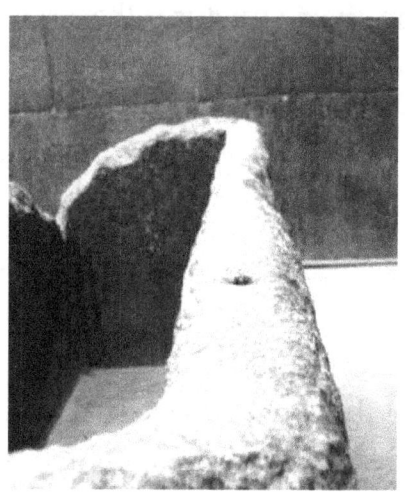

As shown in the previous picture, even the lip of the "vibration cavity" was possibly rounded smooth to insure optimum sustainment of oscillation. This is no sarcophagus. Why in the world would the sides have rounded tops so that a non-existent lid could never stay secure??

Gas to Restart the Pyramid

All that was fine if you only wanted one burst of electricity, but sustainment was certainly needed and it was also planned for. Hydrogen gas was, most likely, produced in what many call the queen's chamber by combinations of materials transferred down the small deposition shafts visible today. Some "acid" would be poured down one shaft and some "hydrated base" went down the other. The volatile gases [hydrogen] produced by the mixture would slowly rise through the baffles [sometimes called the Grand Gallery] and finally reach the "King's chamber" before the "Granite Crystal Oscillations" stopped. Sparks associated with the electricity produced by the granite crystals would ignite the gases and cause the chamber to momentarily get larger from the explosion. This action relieved the pressure on the Granite slabs, which, in turn, produced more cycles of electricity. Very quickly the pressure of the pyramidic weight would take over and begin to crush the granite slabs once again to produce more cycles of electricity over again. Like an almost imperceptibly moving engine, the electricity would continue to be produced as long as hydrogen gas was allowed to enter the "Resonating Chamber". I don't mean small amounts of electricity either. It produced large amounts of electricity. I could talk about the getting bigger, getting smaller, getting bigger, getting smaller reaction ad the explosions and compressions kept occurring over and over again at 640Hertz, but I won't because it would be very boring.

Explosion Evidence

Researchers have discovered that many of the boulders that make up the king's chamber have been moved out slightly and one of the granite slabs has even cracked due to the effects of these cyclic explosions so there is good evidence to support the oscillating cavity theory.

Did I mention that the bottom of the lowest granite ceiling slabs is covered with a fine black dust as if some type of high rate burning process had blackened it? This blackening is found nowhere else in the pyramid. Someone might think explosions occurred in the "King's Chamber".

Starting the Pyramid

After the "queen's chamber" had been filled with the acid and base mixture, and the hydrogen gas had tunneled through the grand gallery to the king's chamber; a flame was introduced through one of the openings to the chamber and the ensuing explosion began the production of electricity and flashing light.

Distribution

Here is the issue. While there is ample evidence to show that the device we call the great pyramid was a compression and tuning set of chambers to produce electricity and there are indications of electricity in use around the world during ancient times, there is a major issue. How was the electric power distributed? For that story we have to learn how to wall through a wall. While it doesn't sound like the same thing, it has great similarity. As a brief overview, let me say that everything in the pyramid was tuned to 640 Hertz. That is, each compression and expansion cycle was accomplished in $1/640^{th}$ of a second. The reason that is significant is that it is believed by some that the earth's base oscillations were to the same timing. The smarty-pants Egyptians [known as the Kemetians at that time] simply made the electricity invisible to the earth and used the earth as a transfer medium to anyone that needed it. There is more to it than this, but we'll get better at understanding how to go through stuff as we go along. The reason that using the dirt to send electricity around instead of wire is that you could pick up the electricity just about anywhere. There was a trick to it because you would have to convert it back to visible electricity, but there were no electric wires needed to transport electricity, so those nasty electricity poles weren't all over the place like we need today..

Where Did The Electricity Go?

The evidence tells us that the "Great Pyramid electric plant" had been built about 40 thousand years ago so you would think that most people around the world would have known about it and there would be evidence of some kind. Well most of the evidence would have been destroyed in that period of time, but there is one oddness that should be considered here. For that we travel to Mexico to a place called Teotihuacan.

While much of the remaining city was built a mere 3 thousand years ago or so, the main two pryamidic buildings in the city were put in place an estimated 8 thousand years ago by a group we can call the preMaya. These first two pyramids were huge.

A comparison of the one known as the Sun Pyramid is shown with the Great pyramid superimposed. Huge pyramids were erected. The two major pyramids were definitely not built by the later Toltec or even the Aztec that followed. The reason this particular pyramidic structure is important to this study is mica.

Mica

Imbedded in the walls and floors of the sun Pyramid was found huge sheets of mica as thick as 1 foot in some areas. Besides being odd, this mica stuff can only be found in Brazil thousands of miles away. I'm not getting into how these people carried huge chunks of mica for thousands of miles to get to Mexico supposedly "without wheels" but you can tell a mystery is brewing.

The scientists were so confused they started looking around some more and in 1971 they found a huge tunnel leading to the pyramid and it had huge chunks of mica as well. Let me tell you what mica is not good for and then we can talk about uses. Mica is a horrible building material as it flakes away. It also would be horrible if one were to line a tunnel used for water as the water would get in between the mica sheets. Whatever was transferred in the tunnel and stored in the Sun Pyramid, it was not liquid. Mica, on the other hand is a great electrical insulator, if one were to find electricity coming out of the ground. One could conjecture that the electricity came from Egypt, but that is just a guess on my part. The electricity could have come from anywhere and the pyramid could have been a conversion and storage device to allow all the comforts of home. A satellite image of the region that has 60 or so pyramidic structures is shown next. The huge one in the middle is this mysterious one and under the long causeway is the mica sheeted tunnel. The huge pyramid at the top of the picture is called the Moon Pyramid and it also was put in place about 8 thousand years ago.

The Great PreMayan Pyramids

Above is another picture of the Sun Pyramid. Only the two largest pyramids stand out as being from around the time of the operation of the Great Pyramid Electrical plant. Besides being the largest, all but those two pyramid structures are covered with artwork and carvings of a later group. Some suggest that the mica sheets that are electricity insulators were used to make the Sun Pyramid and electricity storage device for this part of the world. I won't get into the anomalous things in the Americas that appear to require electricity, but possibly you could get it when you travel to Mexico.

The Pyramid to the Moon is clearly shown in the following picture as the 2nd largest pyramid in the city. Most of the buildings were added at the later date, but the central buildings were here well before the latter group. As an aside, let me tell you a little more about the Aztec that conquered the remaining preMaya. Aztec were really the descendants of the Phoenician invaders. If you recall, the

Phoenicians, by this time, were made up largely of Jewish descendants. So here we have those that escaped the worldwide floodwaters who settled in America only to be overthrown by the Jewish, "Chosen Ones", a few thousand years later. By the way, the Aztec didn't call themselves Aztec. They thought they were the "Mexica" or "Chosen Ones".

We don't know exactly what the Pyramid to the Moon was used for, but we do know that ancient Americans had electricity to electroplate, provide lighting, produce usable titanium, and all sorts of things. With no electrical plants in the Americas, as far as we know, these guys seemed to have been storing reconstituted Egyptian Electricity that had traveled underground or in the air for thousands of miles without wires. If you are in disbelief about electricity going through the ground, wait till we talk about Tesla.

I know you are wondering about how this will help you walk through a wall, but please hang in there. The idea that electricity was common back in the very ancient times required an understanding of relativistic science. Without this understanding, the electricity would not have gotten to Mexico. If it couldn't get to Mexico, the Mica insulation doesn't make sense.

Why 8 Thousand?

I guess you are wondering where I got the 8 to 10 thousand year date rather than the 3 thousand year date you have been told. Luckily, we can date at least one of the pyramid structures in Mexico. For the dating we look at a volcano. The name of the volcanic mountain is Xitli and it has erupted several times in the ancient past. Researchers have found a pyramid that is now called Cuicuilco. Its claim to fame is that it was caught in a lava flow. Lava from Xitli flowed over the pyramid as deep as 20 feet in some areas, but the pyramid was not damaged. It was not un-scarred because ancients had found some newfangled, heat resistant, building materials. The pyramid was so old that it had a layer of soil on top of it thick enough to protect it from the eruptions. Estimates of the eruption dates show that this pyramid **was already ancient over 8 thousand**

years ago. Now that that's off my chest, let's get back to Egypt and discuss something rather strange.

Dendera Tube

I'm still talking about electricity instead of walking through a wall, but this is important. What do Apes, Egypt, and Electricity have in common? The answer or at least the question is found on pictures in a temple in Dendera, Egypt as shown below. These pictures depict something people like to call the Dendera tube. Two of these devices are shown below. The devices appear to be electronic and apes typically are depicted with them.

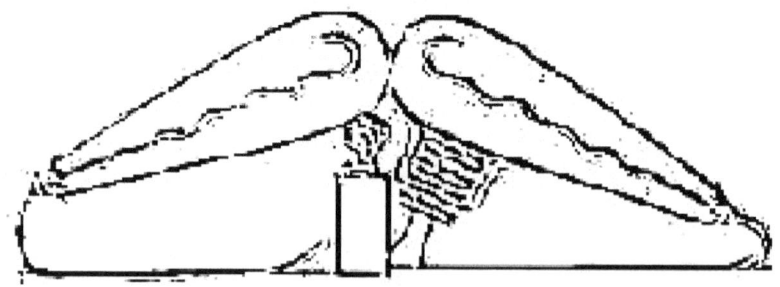

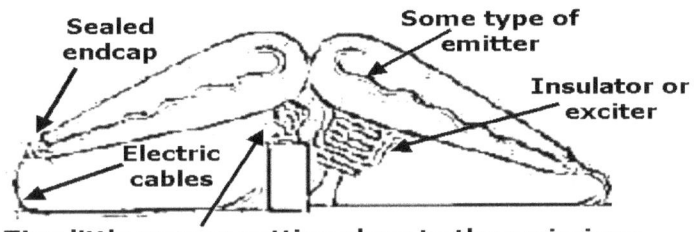

Sealed endcap · Some type of emitter · Insulator or exciter · Electric cables · Tiny little person getting close to the emissions

No one knows what they were for. <u>One possibility was to allow someone to walk through a wall</u>. Some records seem to indicate that these mysterious electrical devices were used to prolog or amplify life. As we go through this book you will see that that life essence has everything to do with walking through walls so let's see where these things were found around the world.

More Tubes

If the first carving were the only depiction, no one would think too much about it, but more and more were found. The following group of pictures following depict some type of electronic tube-like objects resting on other strange devices. A little human who tries to point the main device at a baboon holds all these up. That's right; a baboon.

I'm not getting into the whole baboon thing in this book, but I will tell you that the ancient Jewish book of Jasher indicated that when the Tower of Babel fell, 1/3 of the people were turned into baboon-like people and the Egyptians were chipping away stone to depict baboon-like people. OK! Forget the Baboon and concentrate on the electronic tube thing.

I said forget the baboon. That is not part of this book.

Dendera Tube and Electricity

I've got to tell you that these "Dendera Tubes", by their very looks, must have used some kind of electricity. The twisted cabling to their base, the filament like internal structure, and the radiation type insulator holding the one on the right below all point to the same conclusion. I presented a case for the Great Pyramid producing electricity for Egypt and other places. This could have been one of the devices that used the valuable resource. Cables are attached to the "Dendera tubes". Another image is shown next.

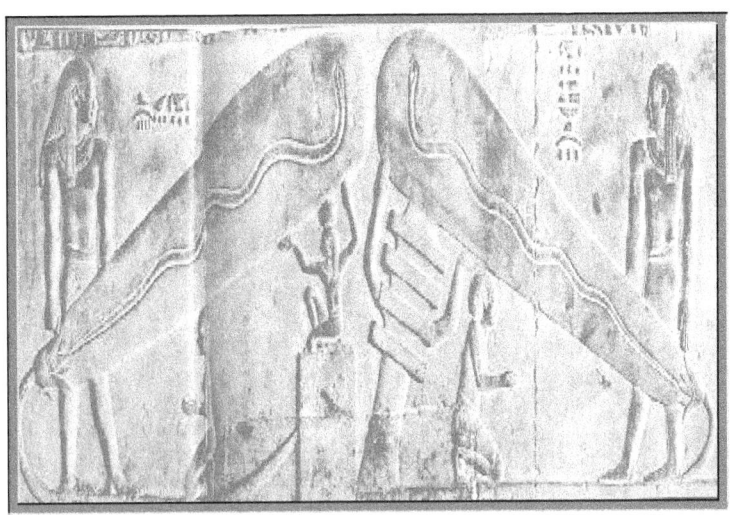

The electricity would have been used to make these things radiate in some way. The radiating component will be explained later, but we should, at least, recognize that the tubes produced something special.

There is something else you should recognize. There is a miniaturized image of Pharaoh under the left Dendera tube on both of the first two carvings. This indicates that the pharaoh needed whatever this machine produced and the pharaoh was insignificant with respect to the Dendera Tube. In the third image following is only different in that the Pharaoh is under the Dendera Tube on the right.

These things must have been really something to be more important than a pharaoh.

The first and third images show a huge baboon guard wielding a knife, evidently to protect the Dendera tubes, so here is my theory, for what it's worth.

Maybe the thing had something to do with keeping the demigod rulers [like pharaoh] alive after the tree of life was lost in the worldwide flood. The Baboons were hired to make sure most people did not live as long as the rulers.

The way one could affect life with some type of radiation is by <u>affecting the vibrational element of life</u> itself. I know you think that DNA structure determines the life cycle, but scientists today know that DNA has no control over such things. Surely, the addition of humans growth hormone has been shown to reduce the effects of age, but the method for its reduction and the reasons why some children die of old age when they are still children has never been explained by DNA. We will get to that later.

As an aside, today, we are using photometric healing devices. These are just high intensity light of various frequencies that somehow rejuvenate our bodies, reduce stress, and increase circulation as those "vibrations" pass though cell to cell to affect our bodies.

More Energy Devices

Sumerian Radiators

Not only do we find these things in Egypt, we find them around the world. The picture below appears to be a similar type of thing and it is being reverenced by 2 regular guys and 2 with wings. I know the wings are curious, but the thing to look at here is the flying ship carrying a person just above the Dendera tube. Versions of this same picture started showing up all over the Middle East during very ancient times. While the tube must have had something to do with flying, there seems to be a strong desire for the "people" in the drawings to be close to this thing as well. Someone concluded that the thing in the middle may have been responsible for keeping the "gods" young. Read the section below from the "Epic of Gilgamesh" and see if this could have been describing the Dendera Tube energy. They are always revered and protected as they were depicted around the world. After all, these babies may have been able to vibrationally change life itself or some equally important function.

Sumerian Epic of Gilgamesh- *Upon her corpse that was hung from the pole, they **directed the pulse and the radiance**; 60 times the water of life, 60 times the food of life, they sprinkled upon it; and Inanna arose*

Sounds like some kind of electric shock treatment, used to bring Inanna back to life. Could the pictured apparatus have been the device used to deliver the directed pulse??

Another Sumerian Radiator

The depiction below is very similar. This time it is men dressed up like fish that are trying to get close to the radiator. On this seal can be seen a tube with some kind of rays emanating away from it. Like others, the device looks very much like the Egyptian Dendera Tubes. We can call this one the "Sumerian Tube". The only difference is that fish men have replaced the baboon and one of those Merkaba is flying overhead. For those not knowing about Merkaba, they were flying transport vehicles. They were described all over the place in the Middle East. I would assume that they were powered by cinnabar and serpent slough like the Indian Vimanas indicated but they also seem to need to "recharge" on top of one of these radiator things.

Another Picture of the Device

Another picture of the ancient Sumerian form of the **Dendera** device is shown next.

Really good Dendera tubes might have been used to help the flying machines rise off the ground by some type of massive levitation thing. Only one fish guy is shown on the above seal, but the depictions are always the same. If you look closely you can see 3 men flying in the thing above the Sumerian Tube.

Possibly Dendera tubes caused levitation and if you could levitate, walking through a wall would be a piece of cake.

Assyrian Radiating Tubes

Here are some examples of the Assyrian version. In this depiction, the radiator seems to have fruit making it more probable that the Nephilim made artificial "Tree-of-Life fruit with the device. A Merkaba [Flying Machine] is still dancing overhead just like in the Sumerian version.

Assyria Fancy Model

Here's one on the next page. This time it's from Assyria. The men have changed clothes again, but the rest is the same. The flying machine has 3 men in it again. The Dendera tube looks a little modern and one of the men has encased himself in some type of shield. Remember this picture when we come to the Indian version.

Assyria Again

In the next Assyrian artifact shown on the next page, the Dendera tube has sprouted flowers and Cherubimic eagle-headed "angels" are picking the fruit. This could have been showing that the Dendera tube was an artificial Tree-of-Life as the eagle gods pick something that looks like fruit from the radiator and there is certainly a reverence to this flowering Dendera tube.

Central American Radiators

The Aztec and Maya both depicted "radiating tubes" of some-kind. Something shoots out of the Aztec version while the Mayan one shows an internal filament similar to the Dendera tubes of Egypt.

Aztec Dendera Tube

The Aztecs must also have known about this device as well. The Codex Nuttal shows the device. Some kind of radiating beam is coming from the central orb. The device is standing by what appears to be a rocket and a throne. No people are in this section of the work, there are no baboons, but a baboon was one of the highest Aztec gods.

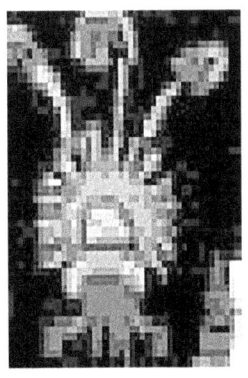

Guatemala Dendera Tube

The Mayans also depicted a radiating tube of some kind as shown in this page of the Dresden Codex. I don't know what the snake thing is in the foreground either. Two filaments can be seen along with the radiating "Glow" around the tube.

Indian Radiator

In India a similar radiating tube was depicted, but this one seems to be somewhat different in that a person gets inside the tube. It certainly was no simple light bulb. [Did you remember the Sumerian guy that also got inside some type of chamber when he was near the Dendera tube thing?]. We can imagine how he got out--- *by walking through the wall*.

Uzbekistan Dendera Tube Inverted

Just like the one found in India, Navai, Uzbekistan was the location of the rock carving shown on the following drawing. Can you see the similarity? Is there someone inside some radiating chamber?

I know there are a couple of odd looking guys holding on to the thing, but there is a person inside. The people outside has things in their mouths so we can believe they are not doing well, but the guy inside is just standing there all happy.

Egyptian Modified Dendera Tube

Almost exactly like the Sumerian version, the Egyptian model changed as shown below. Notice the <u>identical flying thing above the radiator</u> and the complete reverence of the gods. I don't see any fruit in this one, but there can be no doubt of its similarity.

What had been a Dendera tube is now something we find everywhere in ancient Egypt we call a Djed, but it looks like it does the same thing as that lightbulb thing did as the gods worshipping the thing indicates something "special".

Lost Power

After many thousands of years, the continuing explosions in the "King's Chamber" took its toll and the Pyramid Generator began to lose its efficiency as the shape of the chamber changed slightly, as some of the structures chipped away, as the whole pyramid became less efficient to operate. Possibly, a small amount of the generating power was still available about 4 thousand years ago when Moses made the "Arc of the Covenant" to expel stored up electricity it had taken in from the power plant. The energy could have just as easily come from God, but this is another thought.

Finally Gone

Most of the people of the world had no concept of electricity 4 thousand years ago as they had "forgotten" the once great sciences from before the great Babel War, 6 thousand years ago. After that time, it would be centuries before we would once again start to understand the simplest concepts of electricity.

No wonder most people could not walk through walls.

One man began thinking about the oddness of electricity and how people were defining atoms. It simply didn't make sense to him. His name was John Keely. He didn't exactly go through walls, but he was pretty interesting just the same.

Modern Discoveries

John Keely

John Ernst Worrell Keely was born on September 3, 1827 and he died on November 18, 1898. His greatness can be well noted in the rediscovery of walking through walls. He was a US inventor from Philadelphia. Remember the Philadelphia for later walking through wall experiments. In the 1880s John Keely possibly came close to the truth about light. During his investigations, he invented numerous devices that <u>seemed to channel vibrational energy</u>.

He claimed to have discovered a new motive power which was originally described as "vaporic" or "etheric" force, and later as an unnamed force based on "vibratory sympathy", by which he produced "interatomic ether" from water and air. His initial work in the Ethereal regions of atomic structures would be sort of our beginning spot as we try to understand how to let our body ease its way through a wall. His Etheric engine is shown below.

Vibrational Sympathy

He refused to reveal the secrets of his inventions and methods. In 1884, his "Vaporic gun" was demonstrated. His description of what it did is very interesting to use. He stated, *"I take water and air, two mediums of different specific gravity, and produce from them by generation an effect under vibrations that liberate from the air and water an inter atomic ether. The energy of this ether is boundless*

55

and can hardly be comprehended. The specific gravity of the ether is about four times lighter than that of hydrogen gas, the lightest gas so far discovered." [*New York Times*, 22 September 1884] Certainly, he had some things wrong, but he was on the right track by investigating the vibrational component of structure. OK! He mostly put a bunch of long words together, but it's the vibration and ether things that are of interest to us here.

He also indicated, *"It is an elaboration of interatomic ether by vibration. The atomic ether vibrates all around the molecules of matter. There is a magnetic force attached to it at the same time, and it assimilates with the molecular atomic aggregations - that is, assimilates with a certain attractive force that it is hard to tell what it is. I call it a __vibratory negative__. It doesn't act like a magnet drawing metals toward it. There is a certain magnetic effect about it that causes it to adhere by vibratory rotation to different forms of matter - that is the molecular, atomic, etheric, and ether-etheric. The impulse is given by metallic impulses, the rotary power that is formed by etheric vibration - that is the force that holds it in position."* [*New York Times*, 7 June 1885]

Vibratory Negative is a very interesting concept. Just think about it. We now know that if two vibrational components are put together out of phase, they disappear. One is the vibratory negative of the other.

This concept is a basic component in learning to walk through walls.

Vibrating Subatomic Substances

Keely's Law of Vibrating Subatomic Substances also adds light to our investigation. *"Atoms are capable of vibrating within themselves at a pitch inversely as the local coefficient of Gravity, and as the atomic volume, directly as the atomic weight, producing the creative force, whose transmissive force is propagated through subatomic solids, liquids, and gases, producing induction and the static effect of magnetism upon other atoms of attraction or repulsion, according to the law of harmonic attraction".* **[He had correctly understood the perpendicularity of gravity and mass vibrations and their affect.]**

Oscillating Subatomic Particles

Keely's Law of Oscillating Subatomic Particles gives us still more information. *"Subatomic particles, oscillating at a uniform pitch produce the "creative force", whose transmissive form, "Gravism", is propagated through more rarefied media, producing the static effect upon all other sub atomic particles, denominated as gravity."*

Vibration Definition

Finally Keely's Definition of Vibration helps us some more. *"Life in its manifestations is vibration. Electricity is vibration. Vibration that is creative is one thing. Vibration that is destructive is another. Both may be from the same source."* **[He somehow discovered that the Spirit force associated with LIFE was affected by the same type of**

vibration as that found in electricity and he had not even read my book on "Vibrational Matter".

Keely's Inventions

John Keely's inventions were endless. He got the big head and started amplifying his results to impress rather than to study and finally he was labeled a charlatan. While this was a sad time, his work should not be discounted as he ventured into a region that could have enlightened many if they had only kept their minds open. We can be fairly certain that John Keely did not know exactly what he was doing, but he kept on doing it just the same. We can believe that he would somehow modify some materials in the process of doing some exotic experiment and he could not repeat the same experiments because he simply didn't know how to get it to happen again. While history has put him down as a charlatan John Keely opened up the idea of the vibrating atomic cloud filled with ether and mystery. Only now are we regaining the concept, notion and definition that is getting us closer to the point where we can walk through walls.

Keely's Inventions

With his science of vibration, Keely invented many devices. Here is a list of the better-known experimental results.

He Exploded water and produced 29,000 PSI, by use of some type of sound wave.

He Disintegrated quartz crystal using acoustics.

He produced rotation by compound sound waves (patented in modern times by Panasonic as ultrasonic motors) an engine was reportedly driven by the flow of Aether into its components.

He tapped into what he called "Aether flows" to run his engines.

He produced a glowing blue light in water using acoustics (now rediscovered as 'sonoluminescence').

He produced a compound "frequency driven" motor that ran from many frequencies.

He demonstrated a pneumatic cannon powered by release and instant expansion of some kind of bizarre plasma vapor.

He demonstrated an acoustic based flying machine that levitated and propelled itself in the presence of government witnesses.

He demonstrated an acoustic microscope capable of viewing into the molecular and atomic interstices of matter.

He demonstrated a spinning globe that was made to rotate with no outside source of power as a demonstration of the Aetheric flows into matter.

Remember the Aether for when we study how Einstein may help us walk through walls. He determined the same thing about matter.

There certainly was "wonder" in the flamboyant things that were done by John Keely. One of those who evidently saw the wonder of his work was living in Croatia. This guy will get us close to walking through a wall. His name was Nikoli Tesla.

Nikoli Tesla

Lucky for us, John Keely was not the last one to look at vibrational matter and the oddness of light and electromagnetic waves. A newcomer named Tesla began to open our eyes some more. Nikola Tesla was born on July 10, 1856 in Smiljan, Lika, which was then part of Croatia. His father, Milutin, was a Serbian Orthodox Priest, but his mother, Djuka Mandic, spurned his inventive spirit as she was an inventor in her own right. She invented a variety of household appliances.

I know that doesn't have much to do with walking through walls, but Tesla was a very important guy. Tesla studied at several universities, but got into the whole electromagnetic fields thing and went to work as an electrical engineer with a telephone company in Budapest in 1881 and <u>I'm certain</u> <u>he studied the works of John Keely in modifying atoms to</u> <u>get work done</u>. In 1883, he privately built a prototype of the induction motor and came to America to work for this guy named Thomas Edison.

Nikoli Tesla reinvented Alternating Current that we know, now, was in use during a very ancient time in earth's history. Just to make things interesting, let me tell you something odd about alternating current. There is no direct exchange of electrons through a material so there is no electronegative potential that can cause work to be done. It makes absolutely no sense that having electricity go back and forth in a wire could make motors run or electronics electronate. After saying, that let me tell you the alternating current does work. It works by momentarily energizing a structure such that it can make atoms get larger and then reducing the size of the atoms. It acts on atoms kind of like a Laser acts with light. The only real difference is how fast the things oscillate. One could say that AC electricity is very close to Photonic electricity. This vibration causes the magnetic fields of the structure to build. The collapse of the magnetic field allows work to be accomplished. It's almost like magic, but we take it for granted every day. The master of this unbelievable secret was Nikoli Tesla who indicated that its discovery magically came to him. An expansion of this idea may help us go through walls.

A Hint

Tesla didn't stop with the unbelievable alternating current idea. Among his discoveries are the fluorescent light, laser beam, wireless communications, wireless transmission of electrical energy, remote control, robotics, Tesla's turbines, vertical takeoff aircraft, and the Tesla Coil. Tesla is also the true father of the radio and the modern electrical transmissions systems. He registered over 700 patents worldwide and even found a way to make **the ground could carry electricity**. [Just like the Egyptians]

Besides Telsa, no one in recent history has been able to use the ground to carry electricity.

The ground has moisture in it and metals of all types. When electric current is pushed through the ground, the ground simply heats up and no electricity is transferred. In Telsa's configuration, no electric wires were needed to carry electricity as most people believe must be required today. He demonstrated to many this new capability and, reportedly, lit large quantities of electric lamps, which were placed at long distances away from his generator. He, somehow, used the earth as the conductor and lit them without wires. The problem with his newfound distribution method was that no one could control who was using the energy, therefore, no Earth transfer systems were ever commercially produced. No one would gain a profit and Tesla's method died with him. Tesla's typical experimentation involved substantial high voltage discharges as shown above. He started noticing changes in things around all the discharges which would get him to link up with some other guys. When he was 81, Tesla stated he had completed a "dynamic theory of gravity". He stated that it was "worked out in all details" and that he hoped to soon give it to the world. The theory was never published; his work disappeared, but some of his cohorts got very serious about how to dynamically change gravity or the essence of matter. This would have included walking through a wall. A group of these smart guys got together and decided to experiment with a ship. They had no idea some would go through a wall.

Irresponsible Testing

The Philadelphia Experiment

Like the Hutchison Experiments [never mind- this comes later], the same ultrahigh frequency material change may have been accomplished in Philadelphia during World War II. The experiment was done aboard the USS Eldridge towards the end of the war, according to several reports.

No one really knows what happened aboard the ship and many simply consider it to be fantasy. The reason for the uncertainty is that people simply can't believe that what happened in Philadelphia on board the USS Eldridge during a most unusual experiment COULD have happened. Well, my friend, the effects witnessed could have happened and they have happened in subsequent experiments. Most of the reports of that experiment can't be 100% verified and initially they seemed fanciful. The reports were of time dilation, and of the ship disappearing and reappearing at a different location. They all seemed more like science fiction. With our new scientific knowledge we might now begin to believe that there could be more to the story than we initially understood. Let's review some of the details as they have come together to date and see how they go along with walking through a wall. We first have to travel back to 1912.

Background

In 1912, a mathematician named David Hilbert developed several different methods and equations for something he called "Hilbert Space". Dr. Hilbert and another man, Dr. John von Neumann, a brilliant mathematician, got together in 1926 to expand the theories of Hilbert Space and time functions. The plot thickens as a Dr. Levinson came along and developed something he called the "Levinson Time Equations". All three of these guys began investigating the subject of invisibility in earnest in the early 1930's at the University of Chicago. One more guy named Doctor rounded out this group of investigators. His name was Dr. John Hutchinson Sr. who served as Dean at this particular time. I'm not getting into what all these guys did, but let's just say they probably worked on scary things. Math was flowing all over the place and unbelievable answers were the beginnings of a new science of time dilation and magnetic field modification of molecular makeup.

Princeton University

By 1933, something called the Institute for Advanced Study was formed at Princeton University. The group doing this "Advanced Study" included Albert Einstein and good old Dr. John von Neumann. It is believed that the invisibility project was transferred to Princeton shortly thereafter. That's when Einstein's unified fields were studied in detail.

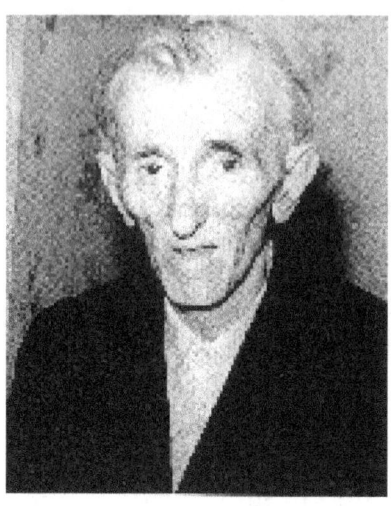

Einstein believed that there was a single particle that made up everything, but one of the offshoots of his theory was that the generation of an incredibly intense electromagnetic field around a huge object like a ship would cause refraction or bending of light and radar waves around the ship. It would be like a massive light refracting bubble. If the motion of the light and electromagnetic waves of radar were wiggled enough, you could not make out the object being viewed. It was like a mirage, where the uneven heating of the air causes the light to bend and wiggle so that objects being viewed become almost invisible over long distances.

Tesla

I'll tell you something right now. You can't do much electromagnetic craziness without getting Nikola Tesla involved to put things into perspective and let you know what the figures mean. Tesla had already invented AC electric generation and efficient manufacture of electricity, some spooky method for transferring signals "Through the Ground" instead of having to use wires, and hundreds of other things involving vibrating electromagnetic fields. In 1936, the secret project was expanded tom include Tesla as the director of the group, shown as he looked in 1942. Yes he did sort of look like a carrot, but it's his brain that was used in this experiment. Together, they studied the nature of relativity and invisibility. You can imagine what the experiments were like, but no record of the details has ever been published or even described outside the tiny group. Some indicate that with Tesla on board, partial invisibility was achieved before the end of the year.

Brooklyn Test

Certainly details are sketchy about this whole super-secret group, but in 1940 a "full" test to make a ship disappear was done in the Brooklyn Naval Yard. Another scientist named, Townsend Brown, evidently, became involved at this point. He had a background in gravity and magnetic mines so he fit right in with the others.

Test Number 2

By 1941, Tesla and his group were given a ship and coils were wrapped around the entire vessel to make it a huge magnetic coil. In some way "Tesla coils" were also used as part of the experiment.

Cancelled

He goes the bad part. Tesla became very uncomfortable because he had a gut feel there would be problems with people being involved in the test. We are told that Tesla somehow knew that the mental state and bodies of the crew would be affected severely. He wanted more time to perfect the experiment but that Von Neumann character thought that there was a reasonable chance at success. Tesla's requests for more time were not answered because the government had a war to win. It is rumored that Tesla went through the motions but secretly sabotaged an experiment in March 1942. Because of this, he was either fired or quit. Tesla died in 7 January 1943 before the fateful test that the "Rainbow Project" is best known.

Rainbow Project

As I mentioned above, at some point in time the studies became called "Project Rainbow". Allegedly, it was an experiment conducted upon a small destroyer escort ship, both in the Philadelphia Naval Yard and at sea. The idea was to make that ship invisible to enemy detection; sort of a super stealth sort of thing. To make the entire ship invisible would be no simple task.

Success

Many indicate that the experiment was a complete success and an utter failure. The ship became invisible all right, but something totally unexpected happened, and those placing the men on board the USS Eldridge wished that they had never done some of Tesla's experiments. According to some, the ship was seen one minute and it disappeared the next. The bad thing is that the disappearing act was not the electromagnetic cloak that had been hoping for. The ship, evidently had teleported to another location rather than being hidden from view.

Bad Experiment

If we piece the story together we get a reasonably clear picture of the strange experiment. On July 22nd, 1943, the power to the Eldridge was turned on, and massive electromagnetic fields started to build up. Just think about huge tesla coils arcing and sparking and that might have been something like the effect, I suppose. A greenish fog slowly close over the ship and concealed it. When the fog dissipated, the *Eldridge* was gone. Everything seemed to have worked well above their expectations and about fifteen minutes later the greenish fog slowly reappeared, and the Eldridge was back.

Bad Stuff

When the Eldridge was boarded, the crew topside was found to be disoriented and nauseous. According to some accounts, the crew, was changed to another group and the equipment was modified to be less powerful.. Navy wanted **radar** invisibility, and this other stuff was pretty scary.

The "Official" Navy Record

The Navy has admitted that the U.S.S. Eldridge took part in an "experiment". The experiment described is somewhat similar, but it does not admit the parts about the invisibility and things. The Naval report indicated that the test involved wrapping wire around the hull of the destroyer in an attempt to cancel out the magnetic fields of the metal on the ship. This is known as degaussing. Unlike the Rainbow Project details obtained since then, the coils embedded in the ship would render the ship "invisible" to underwater magnetic mines that rely on proximity sensors to trigger the detonation. These sensors operated by detecting magnetic fields around ships. Without the magnetic field, the ship would be able to pass through regions mined with these sensors, invisible to enemy mines, but not to radar or vision. Let me tell you that the Navy does not always tell the whole story, but it is interesting that the admitted part included building a specially designed ship with a specially designed magnetic field generator to cause a level of invisibility. Let's see where the ship full of coils went next.

Worse Experiment

In October in 1943, a 2nd test on the *Eldridge* was performed. The electromagnetic field generators were turned on again, and the *Eldridge* became near-invisible. It was stated that only a faint outline of the hull remained visible in the water, but then the Eldridge vanished again. This time it was even stranger. Within a few seconds it reappeared in Norfolk Virginia for a few minutes and disappeared only to reappear back at Philadelphia Naval Yard.

While the whole thing was pretty much Top Secret, it is said that the last experiment was accomplished next to the Merchant Marine ship S.S. Andrew Furuseth and one of its crewmen. Carlos Allende, wrote several strange letters to a Dr. Jessup, in the 1950's. He was a tattletale for sure but there were a few other leaks to embellish what would become one of the most amazing tales ever told. The question still asked today is "How much of the story is factual and how much was embellished beyond the facts.

Bad Stuff Again

After the last test most of the sailors were violently sick. Those were the lucky ones. Some of the crew were simply "missing". Some went crazy. A few of the crewmen had done what this book is about.

They went partially through a wall or deck. They became sort of "fused" with the ship itself.

While this proved that people could indeed walk through walls, it was at a terrible price and absolutely no one knew what had really happened. All 181 crewmen were either discharged with medical issues, had disappeared or had died because of the experiment.

It has been told that 2 guys, a Duncan Cameron Jr. and his brother, Edward, were in the control room to operate it. When everyone else had been killed or generally gone crazy, these two guys were shielded in the protected room. As they witnessed things falling apart, began shutting everything down, but it was too late.

Some type of cover up ensued and the crazy sailors were placed in medical care as the United States went on to develop and deliver the Atomic Bombs and the outcome of this experiment almost disappeared from history.

What they had found is that the make-up of atoms is very tenuous at best. It doesn't take much to disrupt and change them a little nudge with high frequency electromagnetic waves completely changed things. It apparently changed time itself for those in the vibrational field. For some they were able to walk through walls.

Hopefully, your interest has peaked, because that experiment was not the last one done. A new inventor uses multiple radio transmitters and high voltage Tesla coils concentrated at a single location to produce "something" and when it is produced the above things occur to objects in this "Field" just like they experienced in the earlier experiment. If you can transform the molecular structure, and tune it so that it can't see your body, you can simply walk right through the wall. In the case of the unfortunate sailors on board the Eldridge, they had no control over the modifications of the decks and walls they were near. The new experimenter is named John Hutchison. This guy has accidentally found the correct vibrations for particular elements to make them appear invisible to one another so that one can pass through another and has made objects appear to be gravitationally invisible to the earth to allow levitation. The vibrations also have mutated the materials to appear to be melted or jelly-like. John Hutchison is one of the miracle workers of this Age. He has discovered, by accident, what man has longed for over the centuries. His rediscoveries include levitation, transmutation, and invisibility. He was able to pass a piece of wood through a piece of metal. The reaction was halted and the wood remained in the metal. Unlike other methods of forcing materials together the resulting combination of materials had no stress signs around the intruding material. Sometimes metals would appear to melt but surrounding wooden objects would not get hot as only the metal structure changed in the field.

Luck or Genius?

John Hutchison

If you haven't heard of this man, shame on America for not teaching about his work going on in Canada. Using combinations of electromagnetic waves, John bombards items to see what happens to them. Objects go nuts. Not to be humble, the effect was named the Hutchison effect. It seems that the effect is not always the same.

Nucleonic Energy

The Hutchison Effect has been determined to be something called Nucleonic Energy as identified and classified by another Canadian named Dr. Mel Winfield. For those interested in the details of his unified particle theory, it's pretty amazing stuff. In his book, "The Science of Acuality" for instance, he defines gravity as the differential velocity between central core elements of a substance with higher rate objects on its outer surface. This effect was actually discovered in 1852 and it was called the Magnus Effect, but Dr. Winfield refined and redefined the elemental parts to allow us to understand the very workings of gravity and levitation. So long as the electrons flew around the nuclear mass at a rate faster than the nucleus, gravity happened and the electrons and, for that matter, the nucleus, stayed together. Each electron banding spins faster than the inner grouping forcing those "electrons" in an orbital around the nucleus. Simply reverse the nucleonic spin and things go flying. Things start levitating. By phasing the oscillations just right, the nucleonic energy is not sensed by the surroundings. It becomes gravitationally invisible.

We are Levitating

Levitation Effect

For John, "sometimes" items levitated when blasted with a wide assortment of electromagnetic frequencies. I'm not talking about a paper clip spinning around on a table by the force of a magnetic field here. I'm talking 50-pound bowling balls and heavy objects. They became "Invisible" to the gravity of the earth because the particles that made up the objects were vibrated just at the right frequency.

While Dr. Winfield was a genius, we cannot ONLY talk about modern levitation and making things invisible without John getting involved. He not only made things invisible to magnetic fields, he made thing completely invisible just like they had done in the Philadelphia Experiment. You have to be a little crazy to do some of these experiments. Remember, John Hutchison is a man experimenting on these dangerous aspects of physics in his own home. John Hutchison has been conducting some amazing, well witnessed, experiments and, I'm sure he will, soon, find out what produces the "Hutchison Effects" that are beyond the initial gravitational invisibility described in "The Science of Actuality". Right now let me just show just a tiny grouping of his levitation successes in the presence of a powerful and strange field of electromagnetic waves. Pliers are yanked into the air, a bowling ball hovers, and various plates fly like planes.

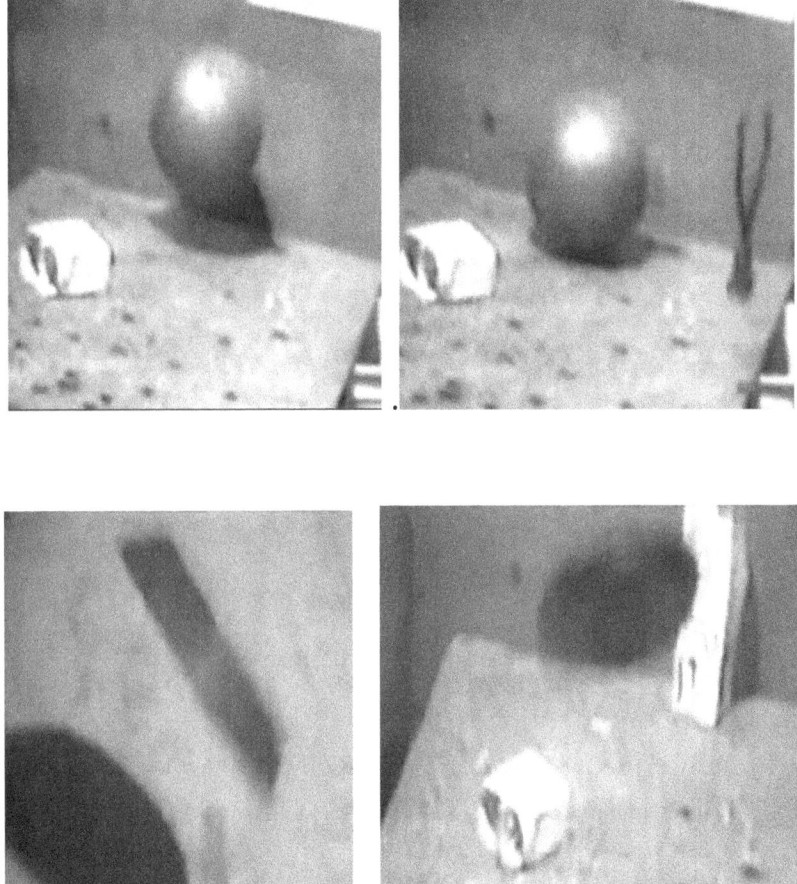

Metal Turns To Jelly

Sometimes metals change consistency and become jelly like before returning to metallic state in a stringy mess as shown next.

 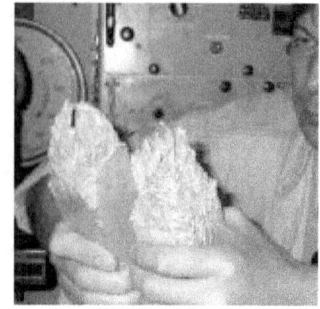

Invisibility

Sometimes objects become begin to fade away and finally they become totally invisible. Below are a couple of images during the transition. No how you can see through the tubes.

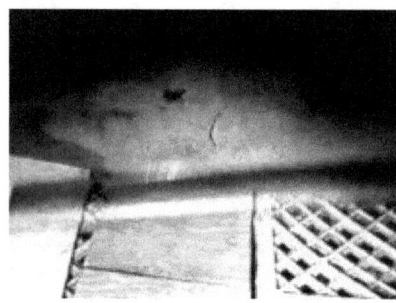 

Objects Go Through Metal

This last phenomenon is the one that provides the most interest with respect to walking through a wall. Sometimes, objects start floating and go right through metal objects. If the process is halted before the material can go all the way through, the object traveling through the metal become part of the metal as shown next. The first picture is of a butter knife that began traveling through a metal plate. It ended up integrated with the metal. It should be noted that the crystalline structure of the metal was not affected by the intrusion. The second picture is of a piece of wood that was caught as it traveled through a plate of aluminum.

John had made thing travel through a metal wall.

John has also had other odd things happen during the experiments, but I wanted to show these to get you thinking that traveling through what we consider to be "solid" [whatever solid means] is very possible and here is the important part. It has been proven. Now all we have to do is figure out what John Hutchison was doing and what he is STILL doing and do it some more. He uses tesla coils, and high power microwave transmitters to concentrate electromagnetic energy on to a small area. Speaking of Tesla, another example might be in order. Let's Walk On Water.

The Bible

Walk On the Water

One way to see if one can walk through a walk is to see if someone can change the characteristics of a material or walk on a liquid. In the walking on water case, the material is again tuned to the body or the body is modified so that water cannot go through the body or be separated by it. For this let's look at ancient texts.

When "God incarnate", Peter, and Elisha walked on water. it did not mean that God made it happen. It means that people can walk on water ---somehow. When Moses, Jannes, and Jambres all made snakes out of sticks for the Egyptian Pharaoh, it didn't mean that God simply made it happen, it means that man can make snakes ----somehow. When these angel characters of the Bible seemed to be able to walk through walls, it doesn't mean that they were going against physics. It means that some elemental form of physics we don't typically use or work was being used to do those things. I'm not getting into making snakes out of sticks in this book; I want to stick to easy stuff first.

I know you are wondering just what I'm talking about on this "attuned to a material" thing. What I mean by that has been studied for many years now. Even without the expansion into the realm of "relativity" which we will have to get to a little later, people began to discover that things are not as they seem. The general study of vibrational differences of materials is called the Tamashii model. This study states that everything—light, shoes, rock, and atoms are simply combinations of vibrational nodes. Get enough of the vibrational episodes together and you can get what appears to be an electron. More combinations seem to make atoms and molecules and structures and everything else.

You are just a bunch of vibration.

Brief Chart of Vibration Matter

Name /atomic #	Maximum Wavelength [meters]	Highest Frequency [Hertz]
Aether		0
Fermion		10^2
Hydrogen/1	1×10^{-9}	30×10^{16}
Berylium/9	1×10^{-10}	30×10^{17}
Silicon/28	3.5×10^{-11}	8.5×10^{18}
Zirconium/91	1×10^{-11}	30×10^{18}
Gold/197	5×10^{-12}	60×10^{18}
Meitnerium/270	3.7×10^{-12}	27×10^{19}
Black Hole		fast

To make this weirder, some matter is invisible.

Vibrational Inversion

If the vibrations of two "objects come in contact with each other and their vibrations are out of phase, they are invisible to each other and they can pass through each other. If you are walking and come to a wall; change its vibrational patterns and simply walk through it because it will be invisible to you and you will be invisible to it like the graviton described below. I knew you would think it's simple!!

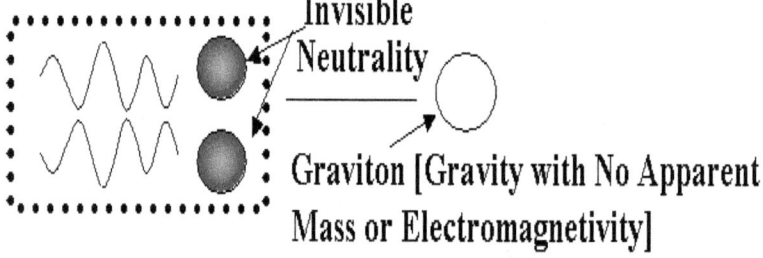

Invisible Neutrality

Graviton [Gravity with No Apparent Mass or Electromagnetivity]

While I'm sure some of you have it completely figured out now, you guys can close the book. For others, like me, we need to get more information. As our first step we need to investigate these "vibrations" the Tamashii models talk about. We'll have to go deeper, but the Tamashii model is a good beginning.

If we could just find matter vibrating in the opposite phase of ourselves we could go there, become invisible, and walk through a wall.

Vibrations

So someone told you that different colors of light were different vibrational frequencies of light and you believed them. Have you ever wondered about these innocent vibrations? To gain insight on light vibrating we must look at some things we "loosely" call atoms. I'm sorry for the confusion you will feel in the next few paragraphs, but it cannot be helped.

I know you have been told all your life that atoms make up everything, but that is pretty much an incorrect thought. While the characterization of the atomic cloud does, in fact, produce things that have similar chemical structure, new science suggests that these atomic clouds themselves are made up of things that don't exist. At least they don't exhibit mass, or nuclear force or some such thing. Sometimes these oddball, quasi-particles are called Fermions, so that is what I'm calling them here.

Experimentation kept breaking down the components of atoms farther and farther. Electrons were made up of quarks, quarks were made up of the fermion things. No one could even begin to imagine something smaller that a quasi-particle. If they are so basic that they must be combined with things to get mass, fermions should be classified as the illusive "Unified Particle" that Einstein was looking for. Unfortunately, they found all types of these fermions. The graviton, for instance has gravity but no mass, even our useful photon has no mass, but it can kill people. There was something more basic.

Fermions Are Not The Answer

While it is impossible to have no mass, these fermion things do the impossible every day and putting enough of these fermions together makes an electron, or a proton mass, or an atomic cloud.

Here is sort of a secret. If ½ of the fermions vibrated out of phase with another half, they cannot make a visible object. They simple make the *"vibrational element"* neutral. Einstein called this characteristic of vibration "Aether", but I think we need to stay with simple vibrations right now.

Don't get comfortable yet. While the fermionic vibrational neutrality is a way of describing invisibility it does not define it.

We can think of these fermions as vibrational energy packets and it's the vibration word we need to concentrate on. Everything seems to be a product of vibration in one way or another, so instead of describing things as mass particles, a better way of looking at everything is by its vibrational essence. Later we will refine this understanding by describing vibrational nodes, but right now just try to imagine all these no-mass things vibrating.

Einstein's Unified Particle

Dr. Mel Winfield wasn't the only one looking for a unified particle to define matter. Einstein also has to be considered. These new fermion things are close to the "unified particle" Einstein was looking for his whole life, but there is enough difference that we will have to discuss him and associated works some more if we are going to walk through a wall efficiently. That will have to wait just a little bit longer as I am trying to get you into this stuff slowly so nothing goes wrong as you move through that wall. Unfortunately, there are all types of fermions, so we have to look at what makes up fermions to find the elusive unified particle. The "people who know" [Don't you hate it when someone says that?] tell us that fermionic differences all come from differences in their vibrational patterns. By changing the vibration, completely different particles can be created or changed.

Wow! We can surmise that the unified particle of matter isn't a particle at all. It is the vibration of nothing or almost nothing. No, I'm going to say "vibration of nothing" is the right answer.

I know this sounds bizarre or stupid right now so don't think I'm offended in any way. Everyone is used to Atoms and structures and matter. Building a world out of vibrations is going to be a job. Making you not believe I'm a nut may take even longer. If you persevere, it will open up your whole life to other possibilities and let you understand things that were impossible to understand before. Let me introduce one of those things right here. Later Einstein will come up with that relativity stuff and others will prove this phenomenon, but right now I only want you to see the problem.

An Anomaly

Two people are together and the same age. One goes traveling at close to the speed of light for 20 years. Neither shaves until the traveler returns. When he returns, the fast guy has not aged a day and the other guy is 20 years older and has a 20-year beard as shown below. It didn't matter where the traveler went and how he came back so long as he did it close to the speed of light. He could even stop every once in a while to see stuff. How can this be? If both are in the same universe, they must be affected by time in the same way.

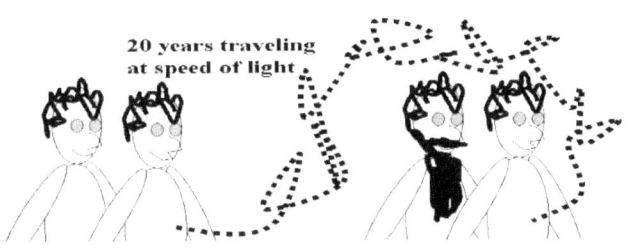

20 years traveling
at speed of light

You may have heard this example in the past, but have you really tried to determine what it means? As we try to uncover a way to go through a wall, some details will be revealed. This is sort of a teaser.

Another Anomaly

Let's take light, or photons, vibrating such that is makes an entire vibrational cycle in 400 nanometers as it whizzes around at the speed of light. The light hits your body and sort of bounces off of it. This isn't exactly what happens, but you get the idea. Light is reflected off your body. Now for some oddness; the photons get mad so they vibrate faster [about 1×10^{-12} meters per cycle.] This time when the light hit your body, it goes right through you and on the way it destroys many of your cells. The beam of light is so happy. Unfortunately you are not in good shape from this "gamma radiation" that you encountered. Why did light change and become so nasty?

Remember this oddness. Light does this nastiness and it has no Mass [most of the time.]

Anthropics

Remember, God Incarnate said one can move a mountain with the "faith" of a grain of Mustard-seed. That does not mean faith in an outcome of your life; it means faith in the SOUL outside of our "Carnal Reality". Well, this means if you want to walk through a wall, you can. Participative Anthropic Theorists have concluded that there is a "substance" that determines our perception of reality. If we had to name this "stuff" we can call it a grouping of souls all attuned to an acceptance of reality. We respond to this accepted view in what we call perception, however, sometimes it is changed for an individual. Let's take, for instance, a small girl picking up a car to help her child or father out from under it so they won't die. The reason you have heard about this is that it happens. I know you have been told adrenaline masses in the body, and puts some shield around the bones so they won't splinter and the muscles turn mighty, but that simply makes no sense. Instead, a person's vibrational resonance can be increased above the level dictated by the mass of souls so that he can affect reality. This thing happens all the time. Books have been written like "Think and Grown Rich" and "The Power of Positive Thinking" which deal in affecting reality by simply knowing something "WILL" happen. We can move mountains if We get really good. Monks in China have been known to lift large stones with their "meditations" to go the "higher chakra" which is the same thing. Don't get me wrong about faith. The faith Jesus was indicating for mountain moving is different that faith in him for the afterlife.

The faith you need for going through a wall is still a real and effective component of our reality that is actually based on something tinier than a fermion.

Oh! I forgot to tell you, the biggest prerequisite for this faith stuff. You must eliminate the desire of self. Even looking too long in a mirror will slam you right back into this "reality" and that stupid wall will be in front of you.

What is Tinier than a Fermion?

The Confusing Part

If adjacent patterns have identical vibrations and are in phase, the fermion comes into existence. It miraculously obtains MASS. Get enough fermions and a boson or a quark appears. A bunch of quarks makes an electron and still more can gather to create an entire atomic cloud. The same things happen with the special fermion we call PHOTON, but something is odd about the photon.

Photons can go right through a wall, so become a photon and you could too.

Fermionic Photon

While there is one school of thought that photonic absorption actually is the methodology for all particles to be generated, that becomes difficult to explain so I think we should stick with another approach and just see why photons go through walls.

Photons go through space fast. In fact they go through space at the speed of light. Einstein and others proved that particles going the speed of light make them have INFINITE MASS.

With this infinite mass, the photon certainly could not go through a wall.

Photons don't have this infinite mass; in fact, they have no mass at all "most of the time". Before I get into the invisibility of light, let me first try to explain the similarity of light and matter. If light can go through a wall and light and matter are the same, we should be able to go through a wall. That brings us to something called Aether.

Aether

Einstein and Dr. Milo Wolff along with many use this concept of Aether when they need to satisfy an equation. Most of the universe is Aether which is smaller than a fermion. While a fermion takes on certain characteristics of matter, but is not total matter, Aether is almost exactly like Electricity in electromagnetic forces. While you think you know what electricity is, you are not right beyond being able to say the word and seeing what happens when it mixes with other things. Electricity is the potential for doing electromagnetic work. It doesn't exactly exist. That is the same for Aether. If you wanted to identify it you might say it is what is between the electrons and protons in an atom. It is the potential for having matter. I thought about this and came up with something I like to call lateral time. If you had millions and millions of Aether or Electricity in a bucket you would have nothing and most of our body is made from the stuff. Aether doesn't combine with anything. It is sort of like a noble gas without the gas part. Anyway a wall is mostly Aether that won't combine with your Aether so you should be able to walk right through. That brings us to a new concept that

Tries to define how God can see and operate on the Beginning and end of time at the "same" time. I call it Lateral time viewing.

Viewing Life From the Side

Lateral Time

For this discussion I will have to pull down this whole concept of time that you currently have. Basically, we can either look at time as a constant, as this impenetrable limit, or as something else. Even in relativistic sciences we must look at light separately from other things because it is sort of and asymptotic limit of matter. In this chapter, I am going to bring out some of the "something" else. We can easily imagine time going forward with the past being behind us and, potentially, we can understand the concept of backward time, but what I need to introduce here is sideways or lateral time. This is not to mess with you, but because several areas in the discussion of light requires a limited understanding of this lateral time and, remember, that light can go through a wall.

Vibrating Time

Before you can really understand time, you must first see it as one of the dimensions that make up what we call a universe. As such, it must be made up of the thing that makes up all things. That thing is not matter at all. It is vibration. I don't mean something is vibrating, I mean that space is vibrating. Today this is called the "Vibrational Motion of Space".

Here is a little secret. This vibration of nothing concept is what makes time, light, matter and life itself. This is some powerful vibrating. Remember, I don't mean atoms or bosons or even the almost unperceivable fermions are vibrating. Time is not made of particles. In fact particles are made of particles.

We accept the vibration of light and the other things around us, but we typically view time as this straight line thing starting at the time we are born and ending when we die. We can, sort of, extend this same "time-line" from the beginning of the BIG BANG thing [if there was such a thing] until the end of all time, but it still is in the same direction and it has the same constant/ linear dimension.

Vibrating Electromagnetics

If electromagnetic fields didn't vibrate, they simply would not exist and neither would light itself.

If we look at atoms, current studies indicate that they are simply clumps of common vibrational nodes rather than true substance.

Sorry I blurted that out. I will explain these node things as we go along. With these new studies we, pretty much, understand that it is the sensing of the vibrations that make these things "known" to us.

If we view time from the side, we can see the beginning of time and the end of time along a line in front of our eyes. I've labeled the viewer as God, because he may be the only one that can perceive this thing. From this vantage point, everything that happens to you from the birth to death are all shown up in one instant. There is no future or past, there simply is.

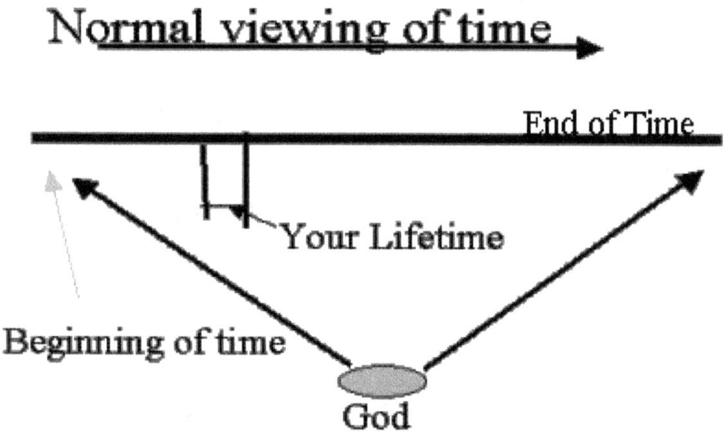

Normal viewing of time →

End of Time

Your Lifetime

Beginning of time

God

As I mentioned at the first of this section, time may be vibrational like all the rest of the dimensional strings of this universe. Instead of a straight line [above], I am showing it vibrating just like everything else does in the following diagram. As we go through time, the humps and valleys don't mean anything to us, but the variations could be witnessed laterally. The hills and valleys might be certain cyclic pressures like destruction periods, Ice Ages, wars, and other things that mark the cyclic nature of time and God could look at all these peaks simultaneously.

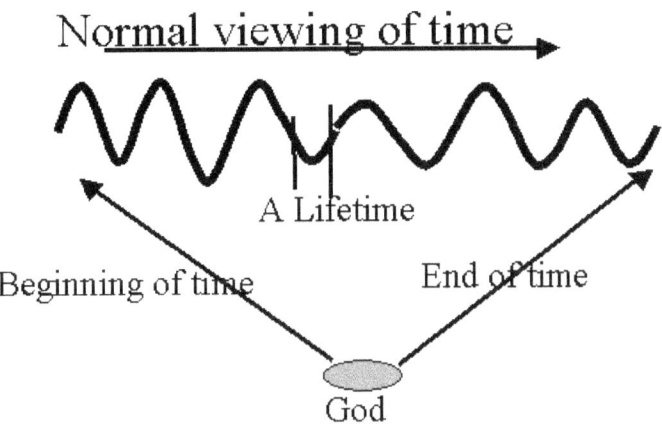

Normal viewing of time →

A Lifetime

Beginning of time

End of time

God

I haven't changed time here; I simply have changed the viewpoint. Notice that a lifetime is shown as a small segment of what would be viewed. The beginning and end of time are only shown for direction. There may be NO beginning and NO end for all we know, so think of the wiggly line going on and on as far as you can see or in a circle as depicted in some examples of dimensional strings that explain the quantum effects noted in life.

Light Seen Laterally

While lifetimes take up a section of the time line, typically light goes the speed of light. It is here one instant and gone the next only to be regenerated and be found again for another instant. Light doesn't actually travel along the time-line. As Einstein predicted, at the speed of light there is no time generated or perceived by the one going that speed. Light or someone going the speed of light would view the universe LATERALLY as shown below.

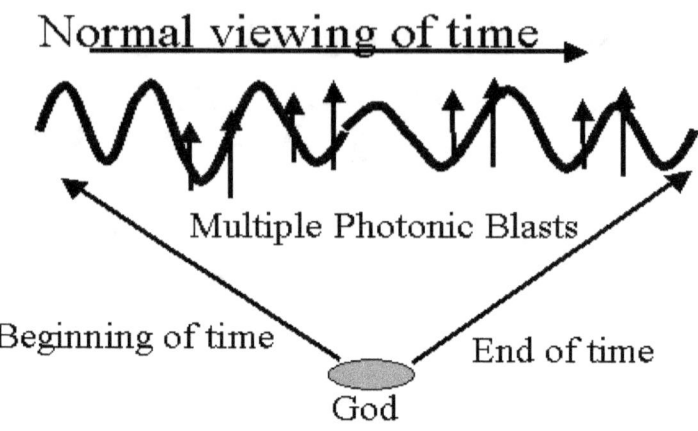

As everything happens instantaneously in this perspective,
Simply figure out how to do this and you could walk
through a wall as it is only there for a short period of time,
and time has no meaning.

Beginnings of Relativity

Let me quickly bring up relativity here. We do not stay on a stationary timeline. If we go in a fast train, perception of time is compressed a little. This happens all the time. If you shine a flashlight while on a train going really fast, how fast will the light be moving to you. Does it slow down by the amount of speed the train is going?---- The answer is no. The speed of light is relative to you. To make up for this variation in perception, we can think of individual timelines for people as going towards lateral time as shown below. When people are going similar velocities, they would perceive time the same way. That is all the people on the fast train would see the flashlight emission as "Normal Light"

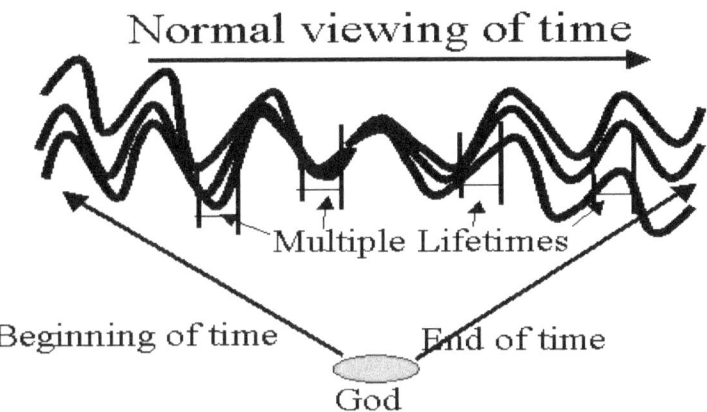

Normal viewing of time

Multiple Lifetimes

Beginning of time

End of time

God

For the odd part about relativity that we will get into later, if you looked at that flashlight emission from the train station as it was shined toward you, it would be going the speed of light, BUT, I vibrational frequency would have slowed down. This is something called "red shift" that I'm not getting into at all. The only reason I brought it up is that you probably heard about this red shift stuff and how Dr. Hubble [of the Hubble Telescope fame] determined that the universe was expanding away from us.

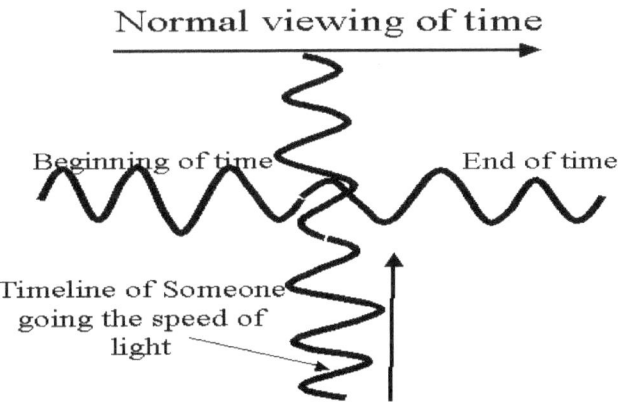

Normal viewing of time

Beginning of time

End of time

Timeline of Someone going the speed of light

Someone going the speed of light would not be traveling along the same timeline. Time for him would be lateral to

101

normal time. While his mass may be infinite, it would be perpendicularly constituted in the normal observer timeline. The speedy guy would experience existence like a photonic blast as shown above. Sure you could go through a wall this way, but there may be other ways that are less damaging to you and won't get you out of sync with the rest of the universe.

Speed of Light Example

That was the easy description with no meat. Let's put on some meat and see what happens. If a person leaves here in a rocket going the speed of light and returns going the speed of light, what would the rocket look like? The answer is that the rocket would gain infinite mass along the direction of travel. It would look like a beam of light and it would not travel on the "normal timeline. The rocket and the person experience LATERAL TIME as shown below.

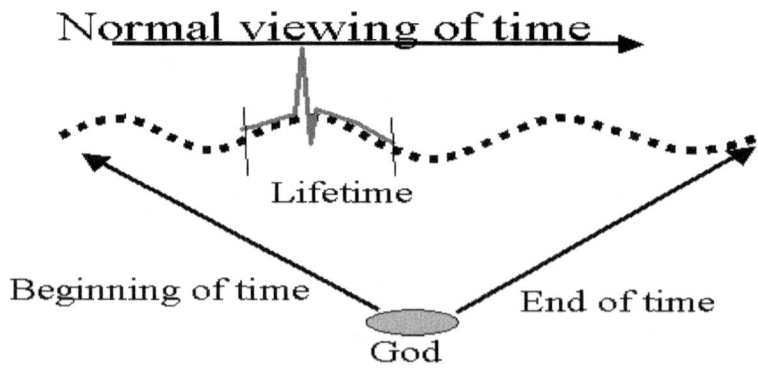

The time-line was expanded a bit to show detail the spike in the middle represents what happens when the rocket goes close to the speed of light. If a person could see what was happening outside, he would see everyone's life passing in an instant to him. The downward portion of the spike is his return to normal home at close to the speed of light.

If you haven't seen it yet, let me tell you that if you go the speed of light you will turn into "light".

Turning Into Light

Let's look deeper. If you could see someone who was viewing you in lateral time, how old would they get as you aged? Of course they would not age a day because they could see your entire life as an instant. If you go the speed of light, how old do you get with respect to those not going with you? The answer is that you would not age. But a man named Lorentz determined something pretty nasty. Mathematically you would not want to go the speed of light. In the Lorentz Transformations matter becomes a squashed ellipsoid with motion. At the speed of light, matter is squashed to it very minimum. It simply is a vibrating line with no width.

A more defining answer is that if you could possibly go the speed of light you become light and you are traveling in lateral time. However, all the particles in your body are vibrating so the particles making up your body are going backwards in time or they must stop vibrating.

While this is one way to address time dilation, there are some others. We will look at other methods later because they truly may help us get through a wall. While high speed is always one possibility, we may not have to go the speed of light to go through a wall.

If you turn into light you can walk through a wall just like other high frequency light.

Let's investigate this vibration just a little more.

More Vibration

Now for the hard one, if everything is really made up of vibrations only, as I have presented in the book on "Vibrational Matter" and this one, what would everything look like when viewed in lateral time? Hurry up; the clock is ticking. -----Come on!

If vibrations are emanations of modification over time, the answer would have to be a solid mass?

This solid mass isn't a mass at all. It is simply a compressed vibration.

Perception

Let's explore this just a minute. How do we perceive the vibrational patterns associated with a mass?

The answer is that we perceive the vibrations as a mass. We cannot see the vibrations and, frankly, we cannot even understand them as vibrations.

It is as if our "impressions" of things are associated with viewing matter in lateral time, where vibrations would be solid' rather than linear time that we experience.

Quantum

I think we need to re-examine the Atom or atomic cloud. The atom is an oddball thing that has specific areas that electrons are allowed and a certain quantity of electrons are allowed in each specific cloud around what we call a nucleus that has a certain number of neutrons and electrons to sustain a certain size atomic structure and diameter. When these "atoms" come into existence, they vibrate at a specific frequency. Changing the frequency changes the characteristics completely. Electrons fling away, protons and neutrons disappear and a new atom appears. Everything is quantized. Even the universe acts this way. I'm so confused I think we need to enlist the help of Einstein and Dr. Wolff.

Albert Einstein

Einstein, sort of helps us some up with a way to go through a wall with something called relativity. Einstein struggled for his whole life with a concept that seemed to never be provable. He determined that the Universe actually was not infinite, but instead it was a 4 dimensional sphere, but he wondered what was beyond his comfortable spherical universe to hold it in place. Now I don't know exactly what a 4 dimensional sphere looks like, but problems soon arose.

While most everyone has heard about Albert Einstein, let me first start with some of his famous and not so famous quotes that are appropriate.

How to solve a problem----*"We can't solve problems by using the same kind of thinking we used when we created them."*

How to be a great athlete--- *"You have to learn the rules of the game. And then you have to play better than anyone else."*

What is a radio------*"You see, wire telegraph is a kind of a very, very long cat. You pull his tail in New York and his head is meowing in Los Angeles. Do you understand this? And radio operates exactly the same way: you send signals here; they receive them there. The only difference is that there is no cat."*

What is common sense----- *"Common sense is the collection of prejudices acquired by age eighteen."*

Thoughts about Education-----*"The only thing that interferes with my learning is my education. Education is what remains after one has forgotten everything he learned in school. Most teachers waste their time by asking questions which are intended to discover what a pupil does not know, whereas the true art of questioning has for its purpose to discover what the pupil knows or is capable of knowing."*

How to be successful------ *"If A is a success in life, then A equals x plus y plus z. Work is x; y is play; and z is keeping your mouth shut."*

What is infinity------*"Two things are infinite: the universe and human stupidity; and I'm not sure about the universe."*

So Einstein was not just an abstract thinker, he actually made sense when he talked. Unfortunately, Einstein himself, worried that his concepts and equations developed over his lifetime might not be correct because of a few little things. So we will look at Einstein and work our way to methods that allow us to, sort of, correct his observations. He and just about all of the main theorists of his day really believed that atomic structure was nodal rather than physical. What I mean by that is ----- Well—Let me just show you what Einstein said.

Time and space and gravitation have no separate existence from matter. *Physical objects are not in space, but these objects are spatially extended. In this way the concept 'empty space' loses its meaning. Since the theory of general relativity implies the representation of physical reality by a continuous field, the concept of particles or material points cannot play a fundamental part, and can only appear as a limited region in space where the field strength / energy density are particularly high.*

OK! This is weird, so let's investigate a little further.

Invisible Vibration Makes Matter

Think about what he is saying. There is no matter per say. The only thing there is in existence is the universe. It's all-together. Certainly things are all around us so let's investigate just what these things might be. Einstein, again, helps us out. He tells us that electromagnetic waves make up everything. In order to give the electromagnetic wave or "field" character, he called the vibrating nothingness the Aether. Let's, again, read what Einstein's own words.

"Since the field exists even in a vacuum, should one conceive of the field as state of a 'carrier', or should it rather be endowed with an independent existence not reducible to anything else? In other words, is there an **'aether'** *which carries the field; the aether being considered in the* **undulatory state***, for example, when it carries light waves? The question has a natural answer: Because one cannot dispense with the field concept, it is preferable not to introduce in addition a carrier with hypothetical properties".*

There Is No Empty Space

Einstein knew that matter only existed as this Aether stuff which was some invisible vibrational or "Undulating State".

Any place that had no Aether could not sustain time itself. Here again is what he had to say.

*There exists an **Aether**. According to the general theory of relativity space without **Aether** is unthinkable; for in such space there not only would be no propagation of light, but also no possibility of existence for standards of space and time,* **nor therefore any space-time intervals in the physical sense**.

Individual Universes

There is one more very important thing to bring up about Einstein and that is what the theory of relativity actually means. It means that the observer defines "time". If someone is traveling close to the speed of light shines a light in front of him, what happens? According to this theory, the light will be going the speed of light to the observer which means that one could say that it will go 2 times the speed of light. Certainly it is not that simple, but. The theory has been tested and proven again and again. Another way of saying this is that each observer has his own universe, so to speak. Let's again look at Albert's own words.

"The second principle, on which the special theory of relativity rests, is the 'principle of constant velocity of light in vacuo.' [to an observer] This principle asserts that light in vacuo always has a definite velocity of propagation **(independent of the state of motion of the observer or of the source of the light).**

Non Relativistic Universe

Many in the physics community still do not try to use the basic elements associated with Einstein's relativistic view of multiple "associated" universes. It doesn't make sense to them so they try to go around it. At a fundamental level modern physicists still try to push through general concepts of particles and fields in space and time.

These people will never be able to go through a wall, because they view the wall as a solid mass.

Even before they can get started they start having huge problems. Unfortunately for them, light and matter exhibit a particle / wave duality. Sometimes they appear to have mass, but not all the time. To explain the duality away, they have built complex models around two main "energy fields". One is called charge (electric and magnetic fields) and the other is called mass (gravitational and inertial fields).

Confusion

Now with a more continuous Electromagnetic Field Theory, "matter" can be interpreted as Particles with 'Charge' OR

Continuous Spherical Electromagnetic Fields. "Light" can be interpreted as Vectored Electromagnetic Waves or the effect of Gravitational Fields.

Back To Einstein

Einstein simply eliminated the duality by ignoring the particles and charges and he stuck with electromagnetic waves and fields. Let's see what he said again. There is little doubt that he eliminated Newtonian physics and began to get mystic.

Physical objects are not in space, but these objects are spatially extended (as vibrational waves or fields). *In this way the concept 'empty space' loses its meaning. ... The field thus becomes an irreducible element of physical description, irreducible in the same sense as the concept of matter in the theory of Newton.*

He continues with some eye opening observations that may help us get through that stubborn wall.

The physical reality of space is represented by a field whose components are continuous functions of four independent variables - the co-ordinates of space and time.

He tries to keep our standard dimensions of length height, width and time, but relativity gets in the way.

Since the theory of general relativity implies the representation of physical reality by a continuous field, the concept of particles or material points cannot play a fundamental part, nor can the concept of motion. The particle can only appear as a limited region in space in

113

which the field strength or the energy density are particularly high.

Dimensions attached to physical shape are dashed to bits and time starts having issues as well. By 1954, Einstein is all but beat as he tried to hold on to a universe with particle defined dimension. Here is what he had to say.

According to the theory of Newton the stellar universe ought to be a finite island in an infinite ocean of space. This conception in itself is not very satisfactory. It is still less satisfactory because it leads to the result that the light emitted by the stars and also individual stars of the stellar system are perpetually passing out into an infinite space, never to return, and without ever again coming into interaction with other objects of nature. Such a finite material universe would be destined to become gradually but systematically impoverished.

"How could this be?" he wondered. In his equations it clearly showed that the universe was a sphere, but if that were so we would be losing energy every day. Depression continued after it was noticed that Dr. Hubble's red-shifts were quantized forming a pretty important observation.

Quantized Red-shifts

Astronomers have confirmed that galaxy red-shifts are quantized. Here is the seemingly weird part. According to Hubble's law, red-shifts are proportional to the distances, galaxies must be grouped into spherical shells, which is odd enough. It gets even odder because everything seems to be concentric around our Milky Way galaxy. It's as if the Earth is the center of the Universe. These shells of matter being around a million light years apart emanating outward from us. It has been stated that the odds for the Earth having such a unique position in the universe by accident are less than one in a trillion.

No Big Bang

Let me tell you what this Earth centered red-shift quantum REALLY means. There could not have been a BIG Bang. Impossible, impossible! If it had occurred, then there would be nothing where the Earth is today. It would be part of the explosion epicenter.

Einstein just could not rationalize the center of the Universe being the Earth. As the great Einstein started doubted everything that he had accomplished, let's see what he had to say.

All these fifty years of conscious brooding have brought me no nearer to the answer to the question, "What are light quanta?" Nowadays every Tom, Dick and Harry thinks he knows it, but he is mistaken. I consider it quite possible that physics cannot be based on the field concept, i.e., on continuous structures. In that case, nothing remains of my entire castle in the air, gravitation theory included, and the rest of modern physics.

I know he sounds like a cry-baby, but he is right. If we want to walk through a wall, we cannot be thinking that particles exist and we have to come to some resolution about this Earth centered Universe along with all the rest of the things that Einstein finally realized as being impossible. I apologize about the cry-baby remark. Einstein was a great thinker and he knew when he had a problem. When he found out, he didn't bury his head.

Dr. Wolff

Dr. Milo Wolff to the rescue. This man really helps us understand what Einstein struggled with. I'm going to go over some of the concepts just a little so that we can be better prepared to walk through a wall. Here is his basic premise.

Perception of Matter is Explained

It is then quite simple to show that: The discrete 'particle' effect of matter is caused by the **Wave-Center of the Spherical Standing Waves** *The discrete 'particle' effect of light is caused by discrete Standing Wave Interactions / Resonant Coupling.*

Dr. Wolff has completely separated the physical characteristics of the universe into NODES or intersections or what he calls "Wave Centers" of these wave things Einstein tried to characterize as Aether [Undulating State]. Just to be clear let me tell you that we are talking about a nothingness vibrating. Where these vibrations cross paths there are areas of "no apparent vibration" [he calls a standing wave]. The standing wave initiates its own characteristic "Out Waves" which makes it appear to have mass. Think of these standing waves are the center of the nucleus of an atomic cloud. I hope that clears it up for you. If not, I'll redirect an answer later. Let's continue.

Time is Defined

Time is caused by Wave Motion (as spherical wave motions of Space which cause matter's activity and the phenomena of time).

Now that the "length, height and depth" things do not stifle us, we can actually define time. I don't mean time on the clock I'm talking about relativistic time where each person has his own observations of time. By all rights we should have introduced the entity of life by now so relativity would mean something, but I'm going to do that later, just understand that the warping of the spherical motions of space are what we can call life. Right now just sense them as these ever-growing spheres of **undulating nothingness** [as Einstein would have said]. Now that time is completely understood [Ha Ha], let's go on to building what we think is mass. Here is Dr. Wolff again.

Forces Are Defined

Forces / Fields are caused by wave interactions of the Spherical In and Out Waves with other matter in the universe which change the location of the Wave-Center (and which we 'see' as a 'force accelerating a particle').

Oh boy, I forgot to tell you about in and out waves. I think I will give you a general concept here. Emanations from inside the universe outward are out-waves. Emanations that

118

push these waves or react with these wave and are initiated outside out universe are the in-waves. While this concept seems simple. It was the major stumbling block for Einstein as he determined that the space outside the spherical universes were completely empty of these undulating. In essence, these in-wave things bring energy into our universe as out-waves continue to leave it. As they are introduced, they affect the characteristic placement of all of the nodes which we characterize as force.

Quantum Defined

Dr. Wolff uses his same omni-observed definitions to show how everything seems to go towards specific quanta.

Quantum Entanglement is likewise caused by the Interaction between the In and Out-Waves and all the other matter in the universe, thus matter is always subtly connected to other matter in the universe (i.e. matter is large not small, we only see the Wave-Center and have been deceived by its 'particle' effect).

Wow! Here we have a true nugget. Matter is not small. What he is saying is that the ends of things aren't really where you believe them to be. They simply have fewer interactions as they emanate to greater distances and cause less and less affect. The quantum effect is actually the effect of the various vibrational waves that case each NODE. Emanations from the node or standing wave must be characterized by the vibrational patterns. Therefore apparent particles would appear at the various vibrational bands of out-waves.

119

Dr. Wolff, continues in his quest of finding definition of the universe by restating Einstein's most important and hardest to understand concept.

However, Einstein's 'Locality' is correct, all matter to matter interactions are limited by the velocity of Waves in Space.

It is "life" that defines "locality". Without life there would be no perception of time and therefore there would be no time. Without time there would be nothing that could define vibration. All characterization of dimension and particles must first be characterized by life itself [Whatever that is.]. Let's think about this red-shift issue that Einstein had. Why should there not be quantized red-shifting rings of mass going outward from the center of a conscious entity's universe?

Life and the Wall

I think we are getting somewhere. All we need to do to go through a wall is to remove life from witnessing the event. OK! It's not a pretty way of going through a wall so we will continue. I think what we need here are 10 dimensions.

10 Dimensional Universe

I have been struggling for a long time concerning the one-dimensional vibration generally defined in a 10 dimensional universe consisting of vibrational matter. Certainly, we should recognize the out-waves and in-waves defined by Dr. Wolff and the relativistic nature of our personal universe as presented by Einstein, but all the standing waves and vibrating nodes still must be able to define a universe. For that we need dimensions. In order to put a structure that that I could understand and still go along with these "string theorists" who indicate that there are at least 10 dimensions. I defined my own and I made sure Einstein's relativity dimension was one to be considered. While string theorist try to tell you that most dimensions are compactified and invisible [whatever that means], Einstein clearly showed that living entities are major characteristics of our universe. As I just reiterated, he generally stated-- without life there is no universe and Dr. Wolff concurred.

Before we get into dimension defining, we first have to put our minds into a Polar coordinated world rather than the normal Cartesian coordinated one. It was easy before all we had to remember was length, height, and width, all Cartesian sectors and all dimensions mutually perpendicular so that they work separately from each other, while being completely locked into one another. I call this characteristic a dimensional dynamo. I'll bring up the dynamo several times in the following explanations, but the elements of each dynamo are all mutually perpendicularly with respect to a polar coordinate system. Everything seems to be radial in our universe and EVERYTHING vibrates. The graphic below sort or shows how three vibrating dimensional Aether could be mutually perpendicular in a polar way.

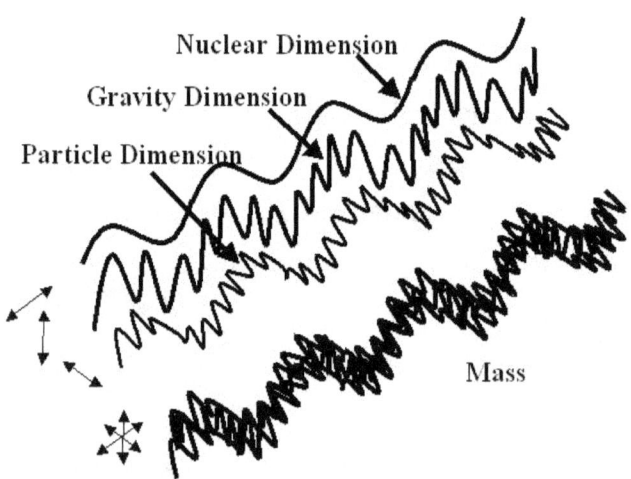

Nuclear Dimension

Gravity Dimension

Particle Dimension

Mass

Perpendicular Dynamos

The 10 dimensions I used to define a universe included 3 mutually perpendicular vibrational dynamos. Each of these dynamos would define an intricate emanation of a universe from its associated vibration and each contains 3 mutually perpendicular dimensions [Similar to the concept of length height, width, but now based on vibrational characteristics.] The first of these dimensional dynamos defines what we can see [or think we see]. Dr. Wolff defined this as the standing waves that initiate out-waves. The second defines how these particles react with one another [motion]. Dr. Wolff defined this function as the aberrations caused when out-waves and in-waves collide. The collisions, he indicated, create force. The third dynamo defines life itself. Both Einstein and Wolff use life experience as an anchor for a universe, but neither tried to describe it very well. I won't disappoint you by defining it very well here, but I think you will get enough of a picture to walk through a wall, because it is the Life dynamo that gives us the most control over our "perceived environment". Each of the dynamos has 3-dimensions. This means that 9 of the 10

123

dimensions are covered by these workhorse dynamos, but we are not finished. A 10th dimension is required to pull everything together and that is the dimension of time.

A general description of the working model of our universe is shown below.

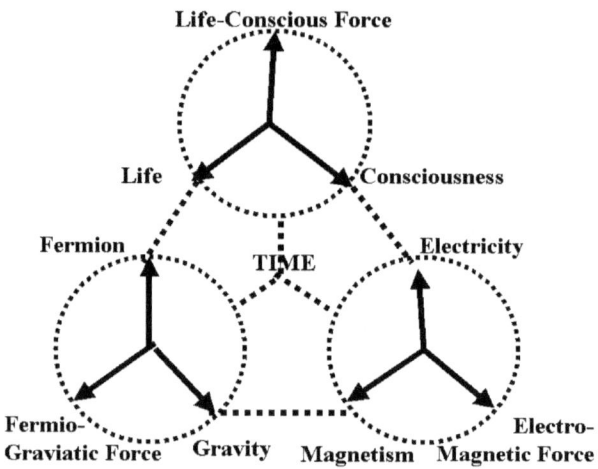

Operational Dynamo

Each of the vibrational dimensions is steered by means of something we call electromagnetism. No one really know what this stuff is, but we can measure it so we use it. In fact, the first dimensional dynamo consists of electricity, magnetism, and photonic force [also called electro-magnetic force]. We know that each of these components are mutually perpendicular by measurements. While we use the characterizations of these dimensions every single day and we even record, measure, and test each of these things, for some reason, no one ever realized that they were required dimensions of a universe. Like all of the dimensional entities these dimensions are just vibrational space. As the electromagnetic fields emanate as in-waves [as described by Dr. Wolff] there are collisions. These collisions are called vibrational nodes or "force".

While the characteristics of electro-magnetics include resonance, and energy coordination, structural dynamos [Matter] can be characterized with the same resonance and energy coordination.

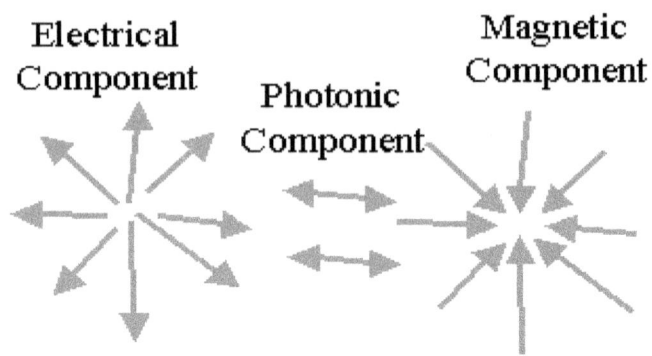

Electrical Component

Photonic Component

Magnetic Component

Electrical Energy

The way electricity reacts is well known in our "normal world". We measure the stuff and use it and it provides energy in accordance with the following equation.

$$E_e = \tfrac{1}{2}\,CV^2$$

This can be considered to be the universal law of electrical dimension, where "C" is the unit of electrical capacitance and "V" is the electrical Storage unit called Voltage. Electrical Capacitance here is the potential to do work in electricity. It is identical to the potential position in the structural dynamo. The Voltage and Velocity elements of the 2 dynamo dualities perform similar actions here as was done in the fermionic equation. As the capacitance increases so does the electrical energy that can be obtained. Voltage, like the velocity affects electrical force by its square.

Magnetic energy has a similar equation with the letter "L" indicating the magnetic inductance. To see how these two things affect the world around it when combined, we find that electro-magnetism resonates or forms NODES in accordance with the following equation.

$$Ro=(CL)^{-1/2}$$

What this is saying is that the highest level of sustainment happens when the electrical capacity and magnetic inductance is at its lowest level. Think of it as electromagnetic entropy. While these equations are used every day, typically we don't see that these equations can characterize ALL operations in the universe. Another important set of equations in electromagnetism is called impedance or reactance. This reactance measurement is the amount of reaction to the environment that is caused by the electric or magnetic field. I'm not going into that here, but let me say that if you can measure that characteristic in one Dynamo, the others can operate and be categorized in the same way. We will look at some of them, but I really am trying to get you to understand how to go through a wall.

Structural Dynamo

As we discussed previously, the vibrational nodes set up their own outwardly bound vibrational waves "particles". The frequency of the waves is dependent on the frequencies of the vibrational nodes, which quantizes the structure of "apparent matter". Higher frequency node carry more waves per second which is characterized as more energy and is perceived to be a mass of more density.

In order to make what appears to be particle, we can anticipate the elements are gravity [inertial differential as described by Dr. Mel Winfield], basic particular matter [perceived as single dimension fermions], and the concept of nuclear force [that somehow holds things together that should not stay together in our non-relativistic, non-vibrational concept of the universe].

Each of these dimensions are required to give the appearance of mass, but none actually has mass. In fact; mass doesn't have mass.

Instead, each is simply vibrating Aether, if put in the words of Einstein. Each dimension exists if the other 2 dimensions of a dynamo exist and any dynamo can exist ONLY if the other 2 dynamos exist. Without the others, the vibrational elements would simply go on forever. The three dimensional elements of the structural dynamo is shown below. Notice that particle components and gravitational components seem to be opposite in affect with the nuclear force holding the particle structure together.

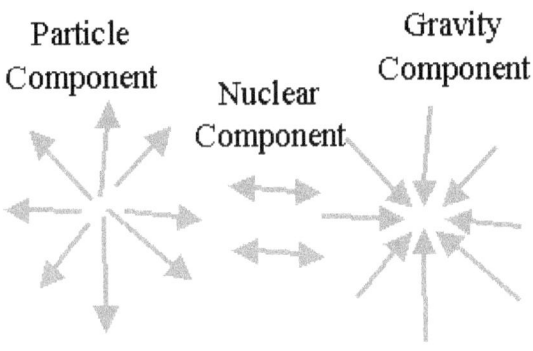

Structural Energy

Like Electro-magnetism, the structural dynamo works to an almost identical set of equations. The energy equation below is the energy produced when an electromagnetic [In-wave] moves a set of standing wave NODES. We sometimes call this kinetic energy, but notice that the energy form factor is identical to those used for electromagnetism.

$$E_f = \tfrac{1}{2} M v^2$$

[Universal Law of Fermionic- or particle dimension] where "v" is the controlling element of Mass [its vibrational amplitude] and "M" is the mass capacity of the system. The anticipated energy from gravity takes on the same general form but we typically call it the energy equation for potential energy.

Particle Resonance

$Rs=(MK)^{-1/2}$ **[universal law of resonance in the structural dynamo]** where R_x is resonance, M is again, the mass capacity and K is the gravitational induction. In this case, the form factor is identical to the resonance equations in an electromagnetic system with the capacity and inductance factors change to those associated with his particular dynamo. Essential, it states that the universe is most affected when kinetic and potential energies are smallest. It is sort of the entropy of particles equation.

Ethereal Dynamo

That brings us to, possibly, the most important and the hardest dynamo to examine or understand. No universe is perpetuated without life. Einstein proved it and many other experiments have proved it since his first conclusions. The Ethereal dynamo must be made up of the 3 mutually perpendicular emanations of life which can be recognized as the living entity, what we can call the soul or consciousness, and a final dimension we can call the spirit [the characterization that allows us to transfer from this universe to a heaven universe.] The last dimension sounds religious, but there is much research that now REQUIRES an adjacent universe to ours. It won't necessarily help you walk through a wall in your present state, but who knows what might happen in a secondary universe.

Some of you simply recognize life as the thing that helps you think and that when your body ceases to function, no life can be enjoyed. For it to have meaning in this universe, some type of linking dimension must connect it. As Einstein and relativistic thinking dictates, this "controlling interface MUST be part of the conscious entity of a universe observer.

I know it sounds like I am some religious nut here, but Einstein presented this problem.

If a tree falls in the woods did it make a sound?

His answer was not only that it did not make a sound, in Einstein's relativity based world, but also it did not even exist until a "cognitive life" experienced the action.

I don't mean a figure of speech here. I mean the tree didn't exist nor the forest until some entity experienced it. Even knowing that this concept has been proven over and over again I had a lot of trouble with the whole relativity thing before attaching a dimensional characterization on life, but we must understand what life means.

Another way to slide through a wall would be to not be cognizant of it when you go through it.

The Meaning of Life

No one knows the meaning of life no matter how many people say that they have found it. Some, who watch too many movies, say it is the number 42. That is not the answer either. While the exact details of what life is, we can be pretty sure that it is CONNECTED to the universe. If matter cannot be created or destroyed, we can assume that LIFE has the same characteristic. I can tell that I losing you. You're thinking that when someone dies, life must be lost, but life may be reborn somewhere else. You are thinking that the number of people continuously increases over time so recycling is absurd, but in fact, if we look at the people living on the earth over the 40 thousand years from Adam

132

on, we would find that there have been many peaks over the ages with greater numbers of people than are in this world today. The peak quantity of conscious lives seems to be regulated by major destruction, war, and catastrophic earth events so that life entities seem to oscillate rather than continually grow as you would think. I have no idea what this means except that we may never truly die.

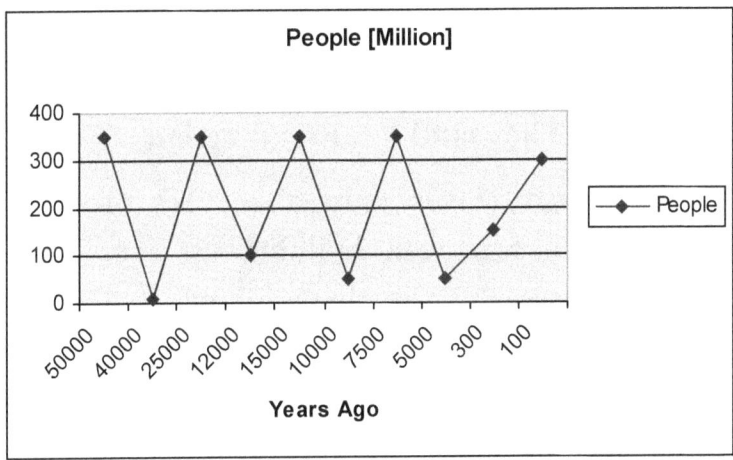

The chart above generally shows what I mean.

40 Thousand Year Correction

There was some type of massive human extinction about 40 thousand years ago. The outcome of this event, apparently, was the emergence of Cro-Magnon humans. Some indicate that this new man started with a guy named Adam.

11 Thousand Year Correction

Another massive destruction occurred about 11 thousand. The ancient book of Jasher indicates that less than 1/3 of all living entities remained.

10 Thousand Year Correction

Again we find a great worldwide flood kills off most of the people as the end of something we call the Pleistocene Age causes massive extinction around the world.

6 Thousand Year Correction

Another huge war leaves only about 1/3 of the world population which has been building up ever since. No telling when the next correction will occur. Could it be 2012 as many seem to predict today? The medieval prophet, Mother Shipton, indicated that it would be 2026 which gives us another few years so I'm using that one right now.

Anyway, all I'm saying is that life seems to be recycled periodically for some reason. With that comforting thought let me get into a few of the nuances with "life" as a dimension and, potentially, bring us a little closer into understanding what life is.

Life Dimension

Let me make something perfectly clear here. When I say life, I don't mean a chain of DNA combined to cause cellular growth characteristics that can grow into an entity. While DNA structure does change the method of cellular growth, there is one thing MUST be understood.

DNA does not make life. Life is a vibrational pattern carried by this Aether stuff that can't be seen, felt touched or understood. Without this life stuff, the DNA might make cells grow, but they would never become alive. Life is a dimensional entity that somehow can be associated with DNA. I know the huge fights that occur when people try to determine the moment that a life starts and simple cell division no longer controls the "life" of a newborn fetus and it would be too confrontational for me to get into that one. I will say again that cell growth and life are totally different. It is the "Life" dimension that controls our ability to do work, to function, to understand, to be self-aware. OK the self-aware thing is a problem if a bacteria is not self-aware or an ameba doesn't know it is living. Let's just say the functioning part of life is the first dimensional component of the ethereal dynamo. Like the other dynamos, the ethereal dynamo works to the same characteristic energy and resonance equations, but I will not get into those in this book. All we really need to understand is enough to walk through a wall and I think I have already gotten you pretty confused.

It is not believed by me that this life dimension is what Einstein and all the rest was referring to when relativity was defined. I think it is the second of the 3 dimensions of the Ethereal dynamo. It is our conscious mind or the thing we typically call our soul.

Soul Dimension

The soul dimension is simply your conscious mind. While you think your life and mind are connected, that is not exactly correct. Certainly both operate in the living entity that is you, but any animal, Tree, bug, or germ can live. The soul is something else. I'm not getting into what animals have or don't have souls in this book, but the idea of being self-aware is the main characteristic. While most people are self-aware, they are not truly aware of themselves. That is they don't recognize the power of the conscious mind.

When God incarnate [Jesus] was on the Earth and told his followers that *if you had faith of a grain of mustard seed you could move mountains* you read that and probably thought----*"That's not me, he was talking about. He is talking about someone who is filled with some type of spiritual entity."* Now we are finding out how very important the conscious mind is for all of reality.

It is the conscious mind that Einstein characterized as the controller of universal existence. That is, the observer determines what is real and what is not.

You have heard stories of levitating people, people who miraculously lifted cars, people who have spontaneously burned up from flames generated by the body, people who have walked on burning coals, people who can heal with a touch, and people who can disappear. ---Forget the last one. I threw it in to see if you were reading everything.

What we will find is that this controller of the individual universe is also the component that will allow us to walk through a wall. Like the other dimensions, the same type of equations as the other dimensions governs the soul.

Soul Energy

Like the other dimensions, the ethereal dynamo works to the same set of specific physical characteristics and it can be defined by equations. Like the others, the "soul/consciousness" energy equation would be of the form below.

$$E = \tfrac{1}{2} QB^2$$

Consider this the universal Law of Conscious dimension. In this equation "Q" is the control of conscious inductance [ability of conscious mind (or soul) to associate itself with its life base and "B" is the Consciousness or that Chakra factor. This consciousness can be viewed as the soul.

$$Re = (QL)^{-1/2}$$

The highest level of life sustainment happens when the life capacity and induced consciousness is at its lowest level. Think of it as life-force entropy. In this case entropy means

138

everything it trying to die or become one with the universe so to speak. We must continuously try to keep ourselves alive or we will be goners. This also means that we have more control over our universe the more we attune our life force energy. Before we get into that, let's first look at some simple tests in vibration. The question is how do vibrations affect our consciousness?

Vibration And The Brain

I know you are struggling with the concept that if you become less conscious, the universe is somehow affected, but the idea that our consciousness is interconnected with the universe is not a new one. Even the concept that vibration controls the level of consciousness is not new. While the concept isn't new, new studies in vibrations and the affects to or by the brain are popping up just about every day. One idea is that if we can lock onto the right vibrational mode, our brains may take over and who knows what might happen.

Let's look at the list of changes in the brain caused or described vibrationally. I'm talking about the alpha, beta, delta, theta and gamma brainwaves, how they affect us, and our ability to walk through walls.

To test and determine what vibrations affected the brain capabilities in what ways, there have been 2 major methods deployed currently. Both of these are audio entrapment processes. That is, they allow the brain to interpret much lower frequencies than those actually transmitted.

Two Testing Methods

The first method binaural interpretation is accomplished by sending slightly different frequencies in each ear. The brain picks up on this difference and uses the difference in its normal excitation.

The second method is called modulation. Similar to the other, this method simply modulates any music or tone very slightly. The brain senses the modulation and interprets that as its control function. Certainly there have been blinking lights, mechanical vibration, and magnetic modulation methods, but the audio methods are extremely easy to accomplish.

The Infratonic Qui Gong Machine, for instance, was developed out of scientific research in Beijing China which studied natural healers and found that most powerful healers were able to emit a strong infrasonic (low frequency sound) signal from their hands. The sound emitted from average individuals was only a hundredth as strong. "Infratonics", is now used by 1% of all doctors in the United States and it is believed that it also is an audio modulating device.

My Experience

I tried one of these Infratonic Subtractors, as I tried to experience separation of the consciousness from the body. After a number of sessions, I was able to completely eliminate the feeling of my legs as I began to hypnotize myself by eliminating the reality around me. My body tingled but one must try not to sense that sort of thing to allow for reasonable separation. After a number of days, I believe I witnessed some other place. All I can tell you is that my body felt like it was vibrating and there was a warmth about the experience. No ugliness, and no thought of self is pretty difficult, but there was a please I cannot explain. I can't remember what I saw, or much of anything else, but some tell me they are able to experience a brand new view of life. I recommend eliminating thoughts of self and even trying the Infratonic subtraction tapes available today. Possibly if you completely lose yourself, you can walk through a wall.

A Dr. Keely [different guy than John Keely] designed something called the Krell Helmet that relied on electromagnetic fields generated in the helmet. By having one side of the brain vibrating at slightly different frequency than the other would have done the same thing as the brain would have subtracted the subsonic vibration as a catalyst to extend the control of the SOUL.. I don't know how successful this machine was, but it illustrates the point that everyone is trying to artificially excite different levels of consciousness and some are beginning to get success.

Oops

Not everyone has gained success as can be illustrated with something called Sphincter Resonance. In the 1960s, somebody discovered the resonating frequency of the sphincter. Presumably, this team created a device called an "Anal Sphincter Resonator". It was supposedly kind of like a musical organ. The idea was to intensify the suspense in movies whenever "Danger" was about to be portrayed. BACKFIRE and more BACKFIRE. Apparently it caused the entire audience to soil themselves. The specific group of tones generated by this contraption has been referred to as a 'Brown Note' for some reason that I am not going into at this time. The specific notes have been lost over time, so I'm sure one of these mishaps will occur again in the future.

The following shows some of the findings that extend beyond what I presented at the beginning of the book. While there are specific frequencies that cause each of these affects, I'm not going to go into that detail in this book. My main objective is to show that vibrational fields greatly affect consciousness.

Type	Freq. (Hz)	Normal Reactions
Epsilon	<0.5	Extraordinary states of consciousness, High states of meditation, Ecstatic states of consciousness, High-level inspiration states, Spiritual insight, Out-of-body experiences, Yogic states of suspended animation.
Delta	0.2 to 4 Hz	Adults are driven to slow wave sleep Confusion, or disorientation Linked to boosting intuition Linked to cultivating "psychic skills" Allows us into "universal knowledge," Access to *external* unconscious material. Deep sleep, trance, Forces irritability Lucid dreaming, intuition, hypnosis Increased immune functions, Decreased awareness of the physical world. Deep dreamless sleep, & Suspended animation, Divine knowledge Anti-aging. Reduces cortisol Increases DHEA & melatonin Empathetic attunement & instinctual insight. Miracle type healing, Inner being & personal growth, Trauma recovery, "One with the universe" experiences Near death experience, Bliss

Type	Freq. (Hz)	Normal Reactions
Theta	4 – 7 Hz	Young children experience drowsiness Adults experience arousal Initiated during daydreaming Can be induced from idle thought Initiate insight from "subconscious mind" Deep relaxation, Meditation, Increased memory, Focus, Creativity, Lucid dreaming, Hypnagogic state, recall, Fantasy, imagery, creativity, Planning, dreaming, Switching thoughts, Zen meditation, drowsiness; Access to subconscious images Deep meditation, Reduced blood pressure Cuts down on mental fatigue, Increases sex drive, Intuitive Augmentation Profound inner peace, Transforming normally held limiting beliefs, Physical & emotional healing, Purpose of life exploration, Inner wisdom, Faith, Some psychic abilities, Retrieving unconscious material, Reverie Bursts of inspiration, Twilight sleep learning, High levels of awareness, Vivid mental imagery. Military remote viewing memories & emotions, Sensations. Induced mental arithmetic Common in extroverts Hypnopompic & Hypnagogic states

Type	Freq. (Hz)	Normal Reactions
Alpha	8 – 12 Hz	Cause or are caused by Relaxation Cause or are caused by Meditation Can be induced by simply closing the eyes. Light relaxation, "super-learning", Positive thinking. Conducive to creative problem solving, Accelerated learning, Mood elevation, Stress reduction, Intuitive insights, Inspiration, motivation, Daydreams Relaxed, yet alert. Calm, relaxed, unfocused, Lucid mental states, Pleasant drifting feelings, Mental resourcefulness, Aids in mental coordination, Amplifies dowsing, Bridge between conscious and subconscious, Non-drowsy but relaxed, Tranquil state of consciousness Body/mind integration. Detachment, Can cause epileptic seizures
Beta	12 – 30 Hz	Associated with alertness Associated with anxious thinking, Can be induced by active concentration Analytical problem solving, Judgment, Decision making, Processing information of world around us. Increased mental ability, Focus, Good for absorbing information passively, Treating Hyperactivity Sensorimotor Rhythm. Problem solving, Used in the treatment of mild autism motivation, Outer awareness, survival, Arousal, Dendrite growth, Combats drowsiness

Type	Freq. (Hz)	Normal Reactions
Gamma	30 – 100 +	Cognitive skill heightened Motor functions heightened Enhances transcendental mental states Boosted memory - 40 Hz Enhanced perception of reality Binding of all senses: smell, touch, etc. Increased compassion, High-level information processing, Natural antidepressant. Improved perception Advanced learning ability Enhanced perception of reality Intelligence (I.Q.) Increase-especially 40 Hz, Positive thoughts & Higher energy levels, Rejuvenation effects Decision making in a fear situation, Muscle tension, Pituitary stimulation to release growth hormone, Helps develop muscle, Recover from injuries,

Let me just give you a few of the frequencies in case you want to try some of this stuff.

Hertz   Condition_____

0.30   Depression

0.40   Confusion

0.50   Relaxing, Also for lower Back pain

0.90   Euphoria

1.00   Feeling of well-being, & harmony

3.50   Feeling of unity with everything

3.60   Remedy for anger & irritability

147

5.80   Reduced Fear & Dizziness

6.26   Confusion, anxiety, depression

6.30   Accelerated learning & increased memory retention

7.83   Anti-mind control, stress tolerance

8.6   Induced sleep, tingling sensations

15.0   Euphoria

I know none of the frequencies actually allowed you to walk through a wall, but what it does show is that consciousness is greatly affected by the vibrational patterns around us. While the military is experimenting with broadcasting subsonic waves to affect brainwaves and enhance the Delta levels [to confuse and put fear in an enemy], many are now trying to tap into meditative states and learning ability by transmitting the 40-Hertz level. What we are finding is that simple stereo speakers may be the best tool to introduce these GAMMA enhancers. A 200-hertz tone is shot into one ear and a 240-hertz sound is transmitted into the other ear. The brain gets both of these frequencies and tries to mix them together to understand the sound. When the 2 frequencies a beat together, the output becomes the difference or 40-Hertz and the brain begins to learn faster.

Brain Vibrations

These studies being conducted around the world describe how sensitive our consciousness is to vibration. The reason is simple. Consciousness is a vibrational dimension and brainwave studies are not the only way to recognize the vibrational characteristics. A second way is something called chakra so let's look at some of these mystical things.

According to the believers and the testing skeptics there are at least seven of these chakras or levels of consciousness.

Root, Chakra--- "consciousness of food or survival"

Sacral Chakra --- "Consciousness of Sex"

Solar Plexus Chakra --- "Consciousness of Self"

Heart Chakra -- "Consciousness of compassion & love"

Throat, Chakra --- "Consciousness of the truth"

Third Eye Chakra --- "Consciousness of our inner being"

Crown, Chakra ---"Consciousness of worlds beyond"

The one we are most interested in is that top one called the crown. They are sort of represented by the diagram below.

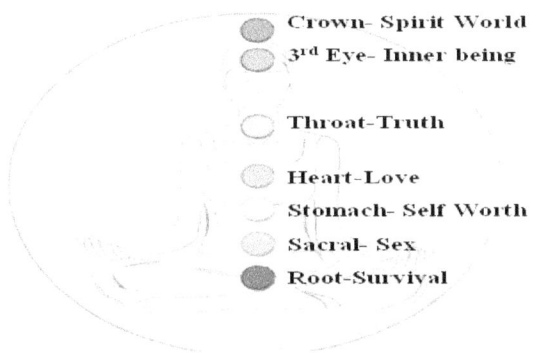

I know it sounds like I'm some guru from India talking about chakras, but it is a convenient way to discuss this dimensional component so I'm going to continue. I'm not putting on the towel on my head, but I may hum a little as I write this section or maybe I'll just stare with my third eye.

The Third Eye

The third eye is derived from a little gland in the brain called the pineal "pinecone" gland. The pineal has no apparent use, but it is thought to have been used by our brains at one time to do all of the things we have been talking about---walking on water, turning one material into another and possibly, even walking through walls. It is thought that the little pinecone used to be a vibration receptor or transmitter. It was a way to affect the world by conscious thought. It certainly would have been used to communicate with animals and other people in the olden days. I know the animal comment got you, but many ancient texts describe that capability and some special people today somehow can get animals to do things no one else can do because of some type of communication no one really knows exactly what it is. No matter what, the gland didn't just grow there for no reason, so let's travel back to the Tower of Babel.

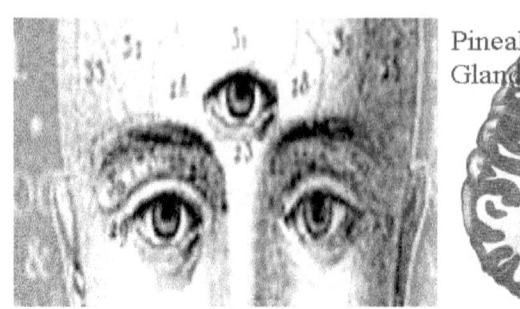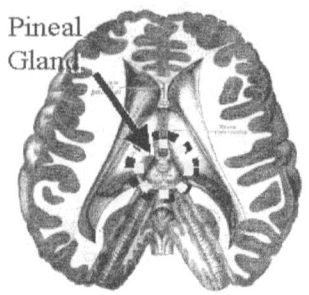

Pineal Gland

Tower of Babel

Most of you know something about this huge Tower that King Nimrod had built about 6 thousand years ago, but many may not know about the huge wars that were written about during this time and how 1/3 of all the people on earth were killed as a result of this war. I could go into what happened when the Tower of Babel was destroyed thousands of years ago and how our brains lost most of their capabilities according to many ancient texts and how these brain losses were probably from some DNA modifying bacteria or something similar. I could also bring out the unusual fact that our current brain size is smaller than our earlier cousins, Neanderthal. While that fact is well known, what is not recognized is that this reduction in brain size shows that our brains began atrophying from disuse about 6 thousand years ago. I could bring out the fact that the entire world was plunged into some type of Stone Age re-insurgence 5 to 6 thousand years ago and people seemed to become dumb as stumps for a while. The Biblical book of Jasher simply tells us that 1/3 of the people died, 1/3 of the people became like apes and 1/3 of the people were dispersed to places around the world because "they could only talk to their close relatives". We can imagine that before this brain reducing started, we could do many things with our bigger brain and operational pineal gland we cannot use today. We can imagine that the pineal gland, prior to whatever happened 6 thousand years ago also was larger and might have been used by our ancestors to affect the vibrational balance of the universe. I could also bring up many other things that would make you wonder if the pineal gland used to allow us to do many things in the past, but I won't. Instead, let me tell you what this tiny, pea-shaped gland does.

Pineal Glands in many non-mammalian vertebrates have a strong resemblance to the photoreceptor cells of the eye. Some evolutionary biologists believe that the pineal cells share a common were the ancestor to retina cells in the eye.

In some animals exposure to light of this gland can change the animal's biorhythm.

Some early vertebrate fossil skulls have a pineal opening so that it probably had some vision characteristic.

The lamprey and the tuatara both have this same type of pineal opening and this thing is photosensitive. The structures appear to include cornea, lens and retina,

The pineal gland is weird in that it has profuse blood flow, second only to the kidney, so we can be sure that it once was of great importance. While doctors are perplexed at why this insignificant gland would need so much blood, it is obvious that whatever happened 6 thousand years ago made the extra blood flow unnecessary.

The brain of a 90 million year old bird was found with a large parietal eye and pineal gland so it's been used for some time now to provide additional insight beyond normal seeing.

Production of melatonin by the pineal gland is stimulated by darkness and inhibited by light. This melatonin stuff affects sex drive.

OK! We have a tiny organ that used to be huge and it used to be an aid in seeing, regulating moods and sex drive, but our bodies are still trying to supply it with enough blood to run a huge organ. Today, the tiny little thing seems to have been abandoned by our bodies, but maybe we just can't see what it can do without vibrating a little.

If we could grow our pineal gland and figure out how to use the thing, we could probably walk through a wall.

Even with a tiny pineal gland, vibrating allows us to understand our world. Abraham Maslow can help, I believe.

Abraham Maslow

This pineal gland/third eye was supposed to have given us the ability to understand the world around us. If we increase our vibrational level by unison with our environment "some call it meditation" or by other exotic means, we can

154

sometimes get in tune with the world around us and here is the odd part. We can even affect it. Another way of saying this is that the 3rd eye thing is that "Self–Actualization" that Abraham Maslow talked about. Self-actualization may help us walk through a wall. The primary component is to eliminate thoughts of self and help as many as you can and you will experience a more complete happiness. That brings us to Norman Vincent Peale and something he called Positive Thinking.

Positive Thinking

Somehow getting our vibrational levels in tune with the vibrational patterns of the elements around us allows us to be more intuitive. We can sense reactions needed to affect the environment. As we affect the environment we can change it. Now the changes are extremely subtle. You cannot, for instance, cause money to fly off a tree, but you can somehow affect the conditions around you that will make it easier to accomplish particular tasks simply by concentrating on these tasks and believing that these things will be accomplished. I know it sounds like gobbly-gook. The problem is that the affect is demonstrated over and over and over again. Positive thinking and getting in tune with the vibrational pattern of the environment actually works. There is no doubt about it. The issue is trying to get into the level of consciousness needed to get the universe to "Bend" a little is not only hard, it also is not easily sustained once one gets to this level of consciousness.

Inductance And The Chakra

Let's think about the conscious energy equation a little more. "Induced Consciousness" would be a reactive component of consciousness if consciousness acts in a similar way to magnetism in the operational dynamo. As we raise our awareness of self, the ability to manipulate an interface between our self and the rest of the universe becomes more defined. The "chakra level" people talk about is simply a vibrational frequency of the conscious. As the frequency increases, the amount we can affect the "life force" increases as well. This is sort of like the magnetism to electricity in the electromagnetic dynamo. Before we get out of this chakra discussion one special chakra level must be explored and this is a wild one.

Crown Consciousness

The answer is held in the last Chakra level. As people expand their awareness more and more or become more self-actualized, as Maslow said, we become more and more conscious of other worlds or universes. Our conscious becomes more in tune with other universes because it is an integral part of this universe. Guess how people describe how to change to higher and higher chakras. They indicate that each expansion is like a vibrational realization. The more the realization or level of chakra becomes; the more noticeable is <u>this strange vibration is felt</u>. The method for getting to this "crown" thing is to sort of hypnotize yourself. "Relax! You are getting sleepy! You are completely unaware of your surroundings! Your feel a warmness and a sense of comfort! Wait just a minute! I'm getting numb here and I can't do the crown chakra dance right now. I have to finish this book.

If you get in this state, the body will begin to hum and vibrate in a low comforting tone and it almost warms your entire body and your feet and hands seem to disappear.

Wow! The first time it happens, the sense of comfort is great but nothing more may have happened. A second and third time might be tried and the world around you is forgotten as you slide into the vibrational level associated with the crown chakra.

Vibration, vibration, vibration. That is all the chakra specialists seem to say. Isn't that the weirdest thing you ever heard? Hold on just one minute. Vibrations are not odd in this book. Vibration frequencies define how a person perceives this universe just like the vibrating fermion that begins to perceive the other particles around it.

As a wall is made to vibrate slightly differently, a person could go right through it.

Let's think of this whole consciousness a little. Let me start over with a question.

Can your consciousness REALLY leave your body?

I'm sure your first thought is that it can't but don't be so ready to close your mind to things that seem to be going on around us. Remember, while consciousness is a special characteristic of each person, it also is one of the dimensions that "build" the particular universe that you experience and it is the various consciousnesses that show the slight differences in our universe. I'm not saying this Einstein is the one that said it and many have proved it.

Consciousness Projection

It is becoming more and more apparent each year that astral projection, near death experiences, seers getting their prophesies, and even reincarnations have been and are elements of the same characterization of the consciousness dimension. I know you think you are using your consciousness right now, but there is more to it that you would like to believe. Let me give you a few examples because, while these are not the same as walking through a wall "physically", they are VERY near the same thing.

Near Death Experiences

It is believed that over 10 million Americans have had Near Death Experiences and lived to tell about it.

One need only go to the near death experience website and find 2000 verified or at least printed events involving near death experiences both good and bad. The website is [http://www.nderf.org] for those interested in a first glimpse. Below is a common theme in just about all of these things.

It is said that the soul [I call it consciousness, but some like to separate us a little more.] goes through this tunnel like "Whoosh".

After the whoosh feeling you are standing in the brightest white light you have ever known. The noonday sun cannot compare to its brightness or stark whiteness.

You instantly feel this bright white light raining down upon your spirit.

You feel an intense love all over your body like soft rain falling on your skin. You know you are loved beyond all shadow of a doubt by this bright whiteness surrounding your spirit.

You feel totally at peace and very safe and love is in you and around everywhere.

You would be typically calm and have no real thought of whatever had "almost" killed you, no pain just absolute peace.

It is said that you feel odd about still thinking, and how alert you are.

Someone may ask, "Do you want to stay or do you want to go back?"

Bad Experience

Unfortunately not all experiences are good ones, however, over 90 percent of the near death experiences are the type mentioned above. Another type is described below.

You feel like you descend – down, down, into a pit, like you'd go down into a well, cavern or cave. And you continue to descend.

Sometimes you can look up and see the lights of the Earth and they finally fade away.

Darkness encompasses you. Darkness that is blacker than any night man has ever seen.

The farther down, the darker it becomes and the hotter it becomes until, finally, you could see fingers of light playing on the wall of darkness.

You feel as if you had reached the bottom of a fire filled pit.

Some see twisted faces grimacing and they are overcome by dread.

Some try to defend themselves against what appear to be ghosts of the damned.

There is a horrible feeling of despair and of being unspeakably alone, abandoned, and suffocating.

Some start screaming and shouting "The heat, the heat!" or "The flames, the flames!" as heard by those witnessing the terrible ordeal. Some die with those words on their lips.

This book isn't getting into hell or anything like that, but it is a very good thing that these experiences are very rare or nobody would want to die.

Near Death Experiences probably are not restricted by walls. Walking through walls might be the least of someone's worries.

I don't know exactly what the experiences show, but one thing is for certain, our conscious mind can go places our bodies cannot. Sometime people just leave their bodies to look around.

Out Of Body Experiences

If we were just looking at out-of-body experiences in general we would find that they are much more common that we would initially believe. This thing occurs in about 1/4 to 1/3 of the population depending on which study you look at. It would be ludicrous to say that up to one third of the human population are mental illness deviants, when in fact, this is such a common phenomenon. They leave their bodies. They see things at great distances from where their body is. They talk to "friendly and informative" people. They insist that the time they are away is not dreamlike, but instead it is close to reality. They usually sense power and freedom. These people recognize and describe objects seen in these states with great accuracy. Others, including many who initially were very skeptical, have verified this strange fact.

Astral Projection

One type of out-of-body experience is call astral projection. Below is a common projection memory.

163

Over several days a possible projector may try to focus on some special place.

Many concentrate on a mantra [some special relaxation word, phrase, or image] prior when falling asleep. [If this seems silly to you try that Infratonic subtraction I mentioned.]

The mantra is used to sort of allow fast self-hypnosis or allow the body to fall asleep faster.

A mantra is said to also help a person stay conscious enough while in a dream state to have control to some level.

 The person would feel their physical body fall asleep while the image of an astral body would emerge and start to rise.

The astral body could go up and out of the experimenter's body.

This is usually easier when the goal was clear.

One can usually feel themselves flying through the air.

As with other out-of-body experiences, sometimes other people can be seen or even talked to during one's travels.

It is said that caution must always be used not to fall into deep dream state at this point.

Many not only view remote sites, but they feel like they gained some personal teaching that stays them in a strong way after the experience.

They, generally, feel at peace when they return to their body.

A Common Thread

Hopefully, you are seeing that in almost all cases, people begin these experiences by blocking out the world including all feeling. Those who are forced in that condition by some tragedy don't seem to have any difference in this the effect. They leave their bodies, get comfort or wisdom, sort of talk to comforting people, get a heightened sense of reality, can float, and when they get focused back on the "real world" they are plummeted back into it. Many times these people are changed forever. The trip to the Crown chakra has changed them forever and I'll tell you why. Their consciousness has been vibrationally enhanced. It vibrates closer to the level needed to do this transfer thing, but in the mean time they become more aware of the feelings of others and become more self-actualized. Whether the "people they interact with are the cause of the vibrational enhancement or some other mechanism is at work, I do not know, but the entire life force of the person is enhanced. Many like it so much they go off and do it again if they can.

There is little doubt that if your consciousness leaves what we refer to as a physical body, you could certainly walk through a wall.

Spirit Dimension

In the ethereal dynamo "light" is characterized by something called the Spirit. In ancient Jewish and Biblical texts, the spirit of being is described as "the light". After the Heaven Wars [The wars where Satan tried to take control of the place called Heaven] the punishment of the rebels was that their "Light" [or spirit] was taken from them. There is a special characterization we need to understand with this spirit thing as well if we are to define light/Spirit more completely. While this particular dimension is one of the neatest ones in the universe, it really does not help us go through a wall until we are "not of this world" so to speak. It is a connecting dimension with a different universe if we are to believe the ancient texts. When God incarnate came to our world, the Bible indicates he left behind the "Holy Spirit". Possibly the union of our spirit and this additional spirit causes some great things that I'm not going into here. Possibly the spirit dimension has a major effect on our universe as well, but I think the "spirit" dimension is too strange and hard to define for now. Also there is the possibility that walking through a wall by some manipulation of this dimension might require dying, so I'm leaving it alone for now. Let's go to the last of the dimensions. I think it can help us a little.

Light Makes Your Universe

When I say light here, I really mean the speed of light, because it is different than the operational dimensions that bring motion. What I'm talking about here is time. The time dimension is certainly the oddest and least understood of all dimensional elements of the universe simply because it is always constant to a conscious entity, but never constant to an outside observer who is in a different motion that the first entity.

Time is so odd. There really isn't a good definition. If we could find out what time was, we could probably walk through a wall for sure.

Difference Between Light and Photons

I know you think that light and photons are the same thing. Potentially they can be associated with one another, but there is a major difference. Light, or more precisely, the speed of light <u>is constant to you</u> in the "reality" around you. I didn't say light was constant except to your consciousness. Einstein found it, but he tried to keep light and photons together and there is where years and years of confusion have been billowing up.

Remember the flashlight shining example? If you shine a flashlight and you are going 1000 miles an hour, the

167

photons will travel from that point at the speed of light [right?] Remember that your mass is slightly elongated and thinned to stay in line with the theory of relativity.

How Fast Are You Going ?

If the 1000 miles an hour worries you, you are in for a shock. If you live in Florida, you are spinning in a circle about 1000 miles an hour already as the Earth rotates. Because the direction is across your body, you are fatter on the equator than at the North Pole where you would go slower. If that isn't enough, we are spinning around the sun at a rate of 66,000 miles per hour. Besides that, our solar system is milling around in the Milky Way at a rate of 43,000 miles per hour.

Now for the bigger numbers the Milky Way is rotating such that it takes 225 million earth years to make a galactic rotation. This means that the galaxy is turning at a rate of 483,000 miles per hour. The picture following sort of shows where we are as we rush along.

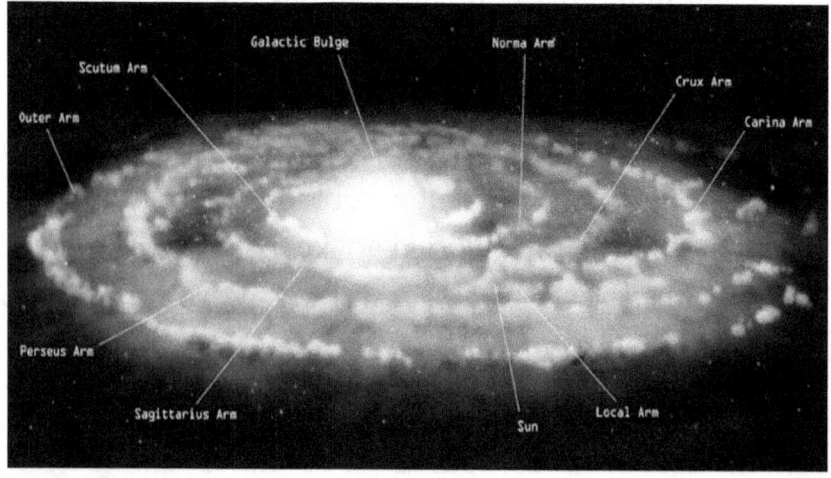

We certainly aren't done yet as the universe appears to be expanding. While expansion is an arbitrary term, red-shift analysis tells us the Milky Way Galaxy is moving at a speed of 1.3 million miles per hour. We are moving roughly in the direction on the sky that is defined by the constellations of Leo and Virgo. The picture below shows where we appear to be going to, currently. I don't know what is so special there, but that is where we are going.

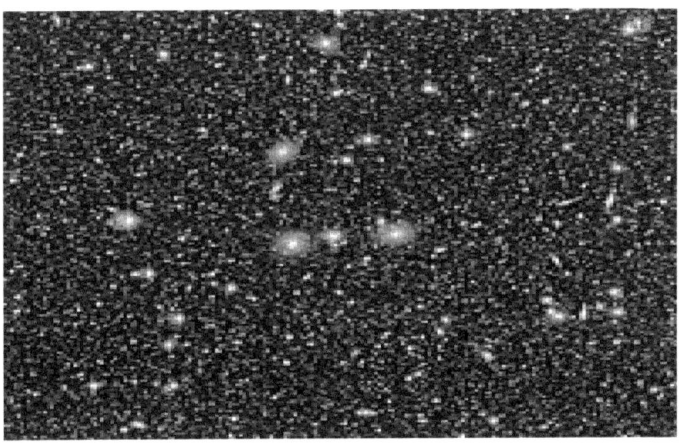

That all being said you are moving at least 1.3 million miles an hour so how fast is light to you? 186 thousand miles per second is always the answer. Someone seeing the light that you are shining that is "stopped" in space sees a light shining at 186 thousand miles an hour as well, but you know in your heart it must be going another 1.3 million miles per hour over that 186,000 miles per second. [It is not?] This is where light and photons separate. Electro-magnetics is constricted by the in-waves and out-waves of the universal sphere. But light stays constant to a conscious entity because consciousness is one of the dimensions of what we can call personal universes.

169

There is a change in the light however. To the stationary observer, the number of cycles a photon vibrates while it is moving is less than experienced by the person on the earth moving so quickly. To the stationary guy, there is a shift towards the slow light oscillation levels or towards the red side of the visible spectrum. It's like the photon stream is stretched to keep both viewpoints seeing light properly in "their" reference point.

Let's Go Faster

Just for kicks, let's see what the light will look like if the difference in speed is really, really high. Let's say the guy shining the light is going close to the speed of light himself. The apparent wavelength would get longer and longer and appear to be radio waves so we couldn't really see a light shining at all.

The only thing to understand is that light is a personal thing in your personal universe, while photons are shared. One way to walk through a wall might be to use this phenomenon. We might be able to walk through a wall that is only referenced in someone else's universe.

Life Is Not Life

Looking at photons is the easy part. Now let's really get into the more exotic thing. That oddball thing is life itself or how time affects our lives. Some just think they go through life at one rate and then they die. Others may understand that life is simply a vibrational dimension, but they have little idea what that means. I went through the "transverse viewing" analogy and showed how, in absolute time, someone experiencing transverse viewing can see all of time in an instant including what will happen in the future.

I believe there are way too many people who have seen glimpses of the future for us to bury our heads in the sand and not believe that there is some way to see the future. I'm not going to get into how someone witnessing future events consciously would necessarily change the future by the action of the conscious mind, because it gets weird. Right now I want to concentrate on walking through walls in the present time.

Let me try something a little different in describing life by using light. While light is constant, we have also determined that there is no aging for someone is going the speed of light to us. It makes no sense to us, but we sort of accept it. Atomic clocks being sent in space ships come home showing the "experienced time" in the space ship was reduced. From these experiments and others, there is little doubt that life is somehow suspended at high velocities.

Unfortunately, it is not a simple observation. If someone was going to the nearest star at very close to the speed of light, he would get there in about 4.5 years. We could watch the event and record the event, but the man in the ship would still not age. His universe slows down as he speeds up, To him almost no time passes. If he shines that flashlight about halfway to his destination, the light from the flashlight would get to the destination before he got there in his world, but about the same time for all other viewers.

This reduction in aging is the best evidence that the Life dimension is associated with individual universes or associated with universes linked by a common velocity. Remember it doesn't matter what direction you move to cause this affect. If you were simply vibrating at the close to the speed of light you would age very slowly and everything around you would simply start rotting before your eyes.

To walk through a wall in this state, simply vibrate until the wall is gone and walk through.

Let's see what we have learned and possibly we can start practicing to go through some walls.

Sadly, I have not been able to get through any yet, but I'm trying.

Going Through a Wall

Let's review some of the ways we can either get through a wall or make the wall vanish.

Peter's Attempt

In the olden days Peter and Elisha did the opposite of going through a wall by not going through water. They walked on water like it was a wall.

Take A Crystal

Moses, Janes and Jambres all took a rod probably with a crystal on the end, made it vibrate and the rod turned into a snake. I know it's not walking through a wall, but it is just as hard to do.

Microwaves

John Hutchison turned on his transmitters and beamed them towards some objects. They flew, they began to disappear and some went through each other as if each was invisible to the other. He never got any volunteers to get in the beam and see if they could go through a wall.

Out of Phase Transfer

Taking 2 fermions with opposite vibrational phases, and their mass disappears so they can go through a wall. Possibly we could find someone that is out of phase with ourselves and get together to walk through a wall.

Fast vibrating Method

Vibrate your body at close to the speed of light and watch the wall disintegrate before you. Then walk on through.

Light Goes Through Walls

By some form of magic, photons simply go through solid objects like walls without effort.

Turn Into Light

Simply going the speed of light allows us to go through a wall just like a photon can go through it. The problem there is that our mass gets really messed up. We probably would not survive such a thing.

Matter Doesn't Exist

By understanding that matter is simply nodes of a vibrational standing wave, we could surmise that bringing in "In-waves" from outside the universe could affect the particles to make them invisible so we could go through a vibrating set of nodes we call a wall.

Tamashii Model

We learned that each material had a special vibrational pattern [60 exahertz made gold]. By simply changing the vibrational pattern to something like water, we should be able to walk right through a watery wall.

Einstein's Observation

Einstein observed that each person had his own universe. His consciousness affected the very nature of his universe. A tree that was not observed didn't exist, so not observing the wall should allow us to walk through it, as it simply does not exist to us.

Transcendental Meditation

By leaving our physical body, or what we characterize as our body we can transcend this state and float through a wall or barrier.

Conscious Transfer

By activating our Pineal Gland or some characteristic that allows us to modify our surroundings, we should be able to walk through a wall.

Infratonics

Try that Infratonic Subtraction to initiate transfer to an alternate reality. Be careful not to find the Sphincter Resonance as you try to go through a wall or you will just "go".

It seems that there are many ways to go through walls. I'm surprised there is even a question about going through walls.

Conclusions, Definitions, Reactions

OK! I wasn't able to give you a simple answer to walking through a wall, but hopefully, you did learn some things along the way that can help you begin your journey. Here are some of the items that you may have learned in this book

- All matter is made up of vibrations not particles.

- Beyond that, everything in the universe can be broken down into vibrations. This includes not only the things we can see like planets and tiny little pebbles, but also electricity, magnetism, photons, nucleons, the spirit, the soul and even life.

- There are 2 types of vibrations. Those emanating from inside the universe going outward [out-waves] and those emanating outside the universe and going inward [in-waves]. Out-waves define matter; in-waves define actions of that matter.

- Instead of three dimensions, there apparently are three-dimensional dynamos, each made up of 3 dimensions to build our universe. None of these can be easily represented in a rectangular quadrant universe we are taught.

- The three dimensional dynamos include the structural dynamo building apparent matter, the operational dimensional dynamo creating force, and the ethereal dynamo creating a life essence that we now know must be present to define a universe.

- An increase in frequency of in-waves and/or out-waves causes more intersections or "standing waves". Each one of these reradiates from what is called the standing wave point which can be considered the center of a new Mass. Higher frequencies, more intersections—more intersections, larger "apparent" atomic structure.

- The structural dynamo is broken down into 3 mutually perpendicular we can call fermionic, gravitational, and nucleatic dimensions.

- The Fermionic Dimension is a characterization of out-waves. It describes the particle component.

- The Gravity dimension is a characterization of these waves and is perpendicular to the fermionic one. It is completely connected to the fermionic dimension and advances gravity into all matter. Increasing frequency increases the effect of gravity. Gravity decreases by the square of the distance between it and its fermionic host.

- When affected by in-waves to establish force, gravity is characterized as a sort of slippage of rotation. That is that apparent inertial characteristics of electron clouds increase in speed as they go away from the central node. The more the inertial difference, the more the apparent gravitation. This expands to any size matter cloud.

- All dimensional dynamos have a connective quality. The Nucleatic dimension is a connective one for the matter producers. It brings matter clouds together in accordance with the frequency base of the fermionic dimension.

- The nucleatic force can sometimes disappear as if it can transfer between 2 universes. In fact, all connective dimensions seem to cross between universes. Besides nucleatic, the connective dimensions are photonic, and spirit dimensions.

- The Operational Dynamo controls motion. It consists of magnetic, electric and photonic dimensions. Each dimension appears to be perpendicular and connected to the other.

- Operational dimensions are created from intersections of out-waves and in-waves.

- The electric dimension builds energy by displacement. It can be associated with potential energy. The higher the frequency, the more electric field is generated as is comes in contact with out-waves electric force is created.

- The magnetic field is associated with the electric field the farther the magnetic field is from the electrical source. When coming in contacts with out-waves, this field also creates a force. The energy level is reduce by the square of that distance from all central nodes.

- Photonic dimension makes vibrational patterns associated with light. Requiring both electrical and magnetic fields, energy increased with the frequencies of the other 2 dimensions.

- Sometimes photons appear to disappear as if they are interface between universes and can transfer back and forth.

- The Ethereal dimensional dynamo is the most difficult to understand it builds life in the universe. It consists of life, soul and spirit dimensions. While they don't seem like dimensions, Einstein and others have proved it to be true.

- The life dimension characterizes matter as living or not living. The level of life is associated with the vibrational frequency of this dimension.

- Associated with life is consciousness. Our consciousness can control our perceptions of our reality including allow us to think through a wall if we believe.

- The last dimension element of our universe besides time is called the Spirit dimension. This dimension restricts and allows transfer of living entities and consciousness between universes. While it is fairly easy to define it is extremely difficult to understand simply because it has no physical characterization beyond the transfer quality. Its necessity is only understood when we die.

- The Time dimension is a personal thing. Separately owned by groups going the same velocity. Changing this can make walls disappear.

- Because of the above, someone traveling really fast ages really slowly. One can even become light. As light, simply making your frequency higher allows you to go through just about anything.

- While light and time are external constants, photon flow is not. As velocity away increases, the photon streams are stretched causing the wavelengths to increase.

- John Keel was a 19th century bag of wind, but he really had discovered the new science that would eventually allow us to walk through walls.

- John Hutchison is able to make things go through other things by changing their characteristic vibrational patterns.

- Einstein became so frustrated with things not adding up that he was saddened the last part of his life.

- Milo Wolff helped clear up some of the issues with Einstein's imperfect relativity.

- Crazy things like chakras and transcendental meditations may help us understand the universe [at least our personal universe.]

The Universe

We must stop trying to fit the universe into the old rectangular coordinated boxes of matter controlled by the size of atomic clouds that were created by some massive heat and pressure at the beginning of time. Nothing supports the notion. Just the idea that light, having no mass can do substantial work tells you this implicitly and everything else around us contribute to the anomaly that is defined in that universe.

Strings

String theorists have, for a long time now, tried to establish an imaginary mathematical model of disjointed dimensions that could allow for the anomalous features, but in order to really adopt the characteristics of our universe, the dimensions have to be completely disjointed and no real structure can be obtained. This is especially true when reaching the speed of light and space-time becomes augmented.

Tamashii Model

The Tamshiists, contend that atoms are made up of vibrating particles. The faster the vibration or the particles, the larger the perceived atomic cloud. The descriptions seem to hold true and they allow for invisibility with in-phase and out of phase vibrational paths, but why the vibrations are here and how the relativistic speeds affect matter are still not clear in such a limited model. If you become invisible you can go through a wall.

Einstein

Einstein threw in a monkey wrench when he set up a relativistic model of the universe where space-time is dependent on the observer. By his day, it was pretty much known that atoms were emanations of waves in some way, but he was indicating that the relative speeds of reactions were dependent on the observer's velocity. He had some conscious control over the universe. And the universe must have some finite structure.

Milo Wolff

Wolff came along and stated that the universe is made up of these in-waves and out-waves that act in a purely polar way as waves emanating from a central node or "Standing way junction of other out-wave form the lattice structure that is to be matter.   Each of these nodal transitions can be considered a center of an atomic cloud [or what we typically represent as a quantized atomic cloud. [Out-waves or the nodal intersections of these polar waves develop structure.]

Quantization

The quantization is directed by intersection. More densely compacted out-waves introduce a higher quantization number, but everything will be quantized to the cyclic nature of these waves that build matter. Quantization demonstrates the wave characteristic of all matter. It is the quantization of velocity vectors that allows us to interact with one another in our vectorized Time universe.

Forces

The joining of in-waves, or operational source, referenced by the standing wave points, or the structural source, controls or establishes the force of each individual node. All motional stresses are caused by the intersections of in and out-waves including [Inertial, kinetic, potential, magnetic, electric, photonic, and nuclear].

Time

The structural and operation universe is completely polar in that all emanations can be characterized spherically. Unfortunately Time seems to act adjacent to the other key elements of matter. While forces all must be identified with time and waves must be characterized with time, we now know that this time component is not constant in an absolute. Time is relative. While one could say that means that the structure of the universe would expand or contract depending on the time constant used for the model. It is this dependency on reality that makes time so hard to introduce. Placement of all of these wave nodes starts shifting and confusion ensues. Time, however, seems to react to velocity. While that seems like a self-fulfilling element as velocity only has meaning with time, the thing to recognize in Einstein's model is that time is "Vectorized" as shown next.

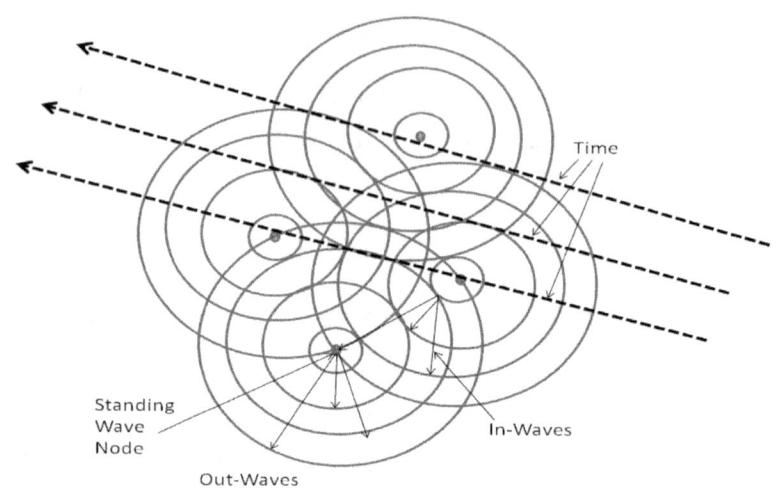

Standing Wave Node

Out-Waves

In-Waves

Time

Vectorized Time

Time stays constant to an observer. Wherever he goes and however fast he goes, "Time" stays with him. That is he experiences everything exactly the same if he goes close to the speed of light or if he could possibly not move through space at all. Experimentally we show that along the axis of a velocity, mass increases. If we could see a table, for instance, the table would get longer in the direction of its velocity vector. NO! NO! NO! In the relative world the table stays the same spatially and the forces [all of the in-waves] react to cause the same forces to the traveler and the objects moving with him. Time can be considered as a rectangular element of the polar universe. Straight-line augmentations of the fabric directed to a vibration. Because none of us are pulling on this fabric the same, the universe must be characterized as multiple universes all being vector driven by time differential associated with time.

Life/Soul

The meaning of life is not 42 but instead it is a complex variant of our universal existence. It is said that if a tree falls in the woods and no one was around, the incidence simply does not happen. Without a cognizant observe, the universe modifications cannot take place. One way to tackle this subject is connecting life to time. That is, time without life or vice versa means little. More precisely, connecting time and the conscious mind [soul] must be done to characterize the universe that changes by perspective in Einstein's relativity theory. They are both undeniably attached and work as a single unit. That is not to say the soul is time because cognition seems to be "timeless" or accomplish adjacent to time. This is saying that the conscious mind is projected into the universe as a vector rather than a polar emission. While in the same direction as time, it is perpendicular to it so it does not place restrictions to time.

Spirit "Energy"

As with all rectangularly coordinated models, if you have 2 mutually perpendicular vectors, there is a third mutually perpendicular vector to balance the other 2. In this case the vector would be insensitive to time or consciousness. Essentially it would be out of this world, but connected to our consciousness directly. It would be able to augment the universe, but the effect might not be readily understood because time was not associated. We can call this dimension the Spirit. This component somehow links this universe and adjacent one or ones. It also is the little voice in the back of your mind instructing you on the way to do what brings the most union with the Universe. Because the vector is perpendicular to the various velocity vectors and abstract to the in and out waves, this spirit dimension is not characterized in the normal sense. You could say it "usually" doesn't exist. String theorists would say the dimension is compactified. I would simply notice that the soul has complete freedom with respect to time. All of the remote viewers and transfer between Heaven and Earth etc. all start making sense.

With that I must end this book. I hope it has been useful to you when trying to interpret what otherwise was considered anomalous in our world. If too many things don't fit, you must try to change your definitions. Consider this as a new definition.

The End

About the Author

Steve Preston is a long time author of scientific, esoteric facts. His series on the creation of mankind is shown below. The series focuses on the painful truths rather than whitewashed details that make us comfortable. If you are interested in the truth instead of comfort, please continue to read and, while you are at it, review other works by Mr. Preston as shown below.

Eight Part Series "History of Mankind"

The First Creation of Man

The Second Creation of Man

The Creation Of Adam And Eve

The Antediluvian War Years

Man After the Flood

A Closer Look At Ancient History

A New View Of Modern History

Other Works

A Closer Look At Genesis

A Closer Look At Lincoln

Adam, Lilith and Eve

America's Civil War Lie

Ancient History of Flying

Behind the Tower of Babel

God Didn't Make The Ape

The Book Of Odd#1

More Oddness

The Day Venus Exploded

The Funny Book of Law

The Truth About Dinosaurs

The Truth About The Earth

Who Really Discovered the Americas?

The Truth About the Heaven War

When Giants Ruled the Earth

Why Are There So Many Anomalies?

The 7 Destructions of the Earth

When Were People on the Moon?

Living on Mars

A Closer Look At Photons

10- Dimensional Universe

Vibrational Matter

Evolution of the Planets

Lizard People

Stupid Science

Investigation In Various Countries

As a serious speculation and history writer, the author has investigated in Israel. He is the guy on the left standing on an embankment in the Negev Desert where the Dead Sea Scrolls were found.

Next, the author rides a camel to the great pyramid of Egypt and travels in a New Zealand cave home of the ancient Maori.

The last image is of the author on the Acropolis investigating the statues of antiquity there.

While this book required only investigation of the mind. The other places were helpful in showing him how amazing things can be.

Thanks for Reading!